Mobile Technology and Place

Routledge Studies in New Media and Cyberculture

1 **Cyberpop**
Digital Lifestyles and
Commodity Culture
Sidney Eve Matrix

2 **The Internet in China**
Cyberspace and Civil Society
Zixue Tai

3 **Racing Cyberculture**
Minoritarian Art and Cultural
Politics on the Internet
Christopher L. McGahan

4 **Decoding Liberation**
The Promise of Free and
Open Source Software
Samir Chopra and Scott D. Dexter

5 **Gaming Cultures and Place in
Asia-Pacific**
*Edited by Larissa Hjorth
and Dean Chan*

6 **Virtual English**
Queer Internets and
Digital Creolization
Jillana B. Enteen

7 **Disability and New Media**
Katie Ellis and Mike Kent

8 **Creating Second Lives**
Community, Identity and
Spatiality as Constructions
of the Virtual
*Edited by Astrid Ensslin
and Eben Muse*

9 **Mobile Technology and Place**
*Edited by Rowan Wilken
and Gerard Goggin*

Mobile Technology and Place

Edited by Rowan Wilken
and Gerard Goggin

Routledge
Taylor & Francis Group
NEW YORK   LONDON

First published 2012
by Routledge
711 Third Avenue, New York, NY 10017

Simultaneously published in the UK
by Routledge
2 Park Square, Milton Park, Abingdon, Oxon OX14 4RN

*Routledge is an imprint of the Taylor & Francis Group,
an informa business*

Library of Congress Cataloging-in-Publication Data
   Mobile technology and place / edited by Rowan Wilken and Gerard Goggin.
      p. cm. — (Routledge studies in new media and cyberculture ; 9)
   Includes bibliographical references and index.
   1. Cell phones. 2. Mobile communication systems. 3. Migration, Internal. I. Wilken, Rowan. II. Goggin, Gerard, 1964–
   TK6564.4.C45.M63 2011
   621.382—dc23
   2011034385

ISBN13: 978-0-415-88955-1 (hbk)
ISBN13: 978-0-203-12755-1 (ebk)

Typeset in Sabon by IBT Global.

Printed and bound in the United States of America on sustainably sourced paper by IBT Global.

Rowan Wilken:
For Lois Wilken, the best aunty you could ever hope for.

Gerard Goggin:
For Jacqueline, como siempre.

Contents

List of Figures and Tables   ix
Acknowledgments   xi

PART I
Theorizing Place and Mobiles

1   Mobilizing Place: Conceptual Currents and Controversies   3
    ROWAN WILKEN AND GERARD GOGGIN

2   The Place of Mobility:
    Technology, Connectivity, and Individualization   26
    JEFF MALPAS

3   Topologies of Human-Mobile Assemblages   39
    RICHARD EK

PART II
Media, Publics, and Place-Making

4   When Urban Public Places Become
    "Hybrid Ecologies": Proximity-based Game
    Encounters in *Dragon Quest 9* in France and Japan   57
    CHRISTIAN LICOPPE AND YORIKO INADA

5   The Urban Dynamics of Net Localities: How Mobile and
    Location-Aware Technologies Are Transforming Places   89
    ERIC GORDON AND ADRIANA DE SOUZA E SILVA

6   The Real Estate of the Trained-Up Self: (Or is this England?)   104
    CAROLINE BASSETT

PART III
Urbanity, Rurality, and the Scene of Mobiles

7   (Putting) Mobile Technologies in Their Place:
    A Geographical Perspective   123
    CHRIS GIBSON, SUSAN LUCKMAN, AND CHRIS BRENNAN-HORLEY

8   Still Mobile: A Case Study on Mobility,
    Home, and Being Away in Shanghai   140
    LARISSA HJORTH

9   Connection and Inspiration:
    Phenomenology, Mobile Communications, Place   157
    IAIN SUTHERLAND

PART IV
Bodies, Screens, and Relations of Place

10 Going Wireless: Disengaging the Ethical Life   175
    EDWARD S. CASEY

11 Parerga of the Third Screen: Mobile Media, Place, and Presence   181
    INGRID RICHARDSON AND ROWAN WILKEN

12 Encoding Place: The Politics of Mobile Location Technologies   198
    GERARD GOGGIN

13 The Infosphere, the Geosphere, and the Mirror:
    The Geomedia-based Normative Renegotiations
    of Body and Place   213
    FRANCESCO LAPENTA

    List of Contributors   227
    Index   233

Figures and Tables

TABLES

4.1   Statistics of the Spatial Distribution of *Surechigai Tsushin*
      Events in *Dragon Quest 9*   62

FIGURES

4.1   The typical range of WiFi-based proximity encounters.   61
4.2   The game terminal interface showing three "encountered"
      players.   65
4.3   A player in the Akihabara meeting spot with the game
      terminal in his right hand and his mobile phone in the other. 67
4.4   Game meeting places.   71
4.5   Connecting to games.   71
4.6   A biker stopping by to play.   72
4.7   The start of the meeting, with the core group of regulars.   75
4.8   Later on with new players, joining in to exchange maps.   75
4.9   Group of players and the cardboard figure marking the
      place.   81
4.10 Players sitting and standing in the group, and a passerby
      walking his dog.   81
4.11 Stone construction on which several players sat.   82
4.12 Group of teenage girls passing by, and a homeless man
      washing his dishes.   82
7.1   Creative practitioners' connections from Darwin to other
      Northern Territory locations, expressed as a percentage of
      all work linkages within and beyond Darwin, 2008.   128

Acknowledgments

The idea for this project emerged over coffee (as all good project ideas do) in a café at the University of New South Wales, Sydney, Australia, in late 2009. This initial meeting, and at least one subsequent meeting, was made possible by the generous support of the Australian Research Council-funded Cultural Research Network (CRN). We are very grateful to the CRN for the collaborative research opportunities this funding has enabled, and we hope that this collection stands as one small example of the many great things that the CRN has made possible.

We wish to express our gratitude to Erica Wetter at Routledge, and the two readers of the collection proposal, for their enthusiastic support of the initial project idea, and to Erica and Felisa Salvago-Keyes at Routledge for helping realize this collection. Thanks also go to our colleagues at Swinburne University of Technology (especially Karen Farquharson), the Journalism and Media Research Centre, University of New South Wales, and the Department of Media and Communications at the University of Sydney.

Our thanks also go to all the excellent contributors to this collection for their faith in this project, their willingness to contribute their work, and for the patience and good humour with which they have responded to our requests, promptings, and questions.

Finally, special thanks go to Karen Olsen for her unstinting support, love, and encouragement, and to Lazarus, Maxim, and Sunday. And to Jacqueline Clark, Liam and Bianca Goggin, for their love, support, and nigh infinite patience.

Rowan Wilken, Swinburne University of Technology
Gerard Goggin, The University of Sydney
July 2011

Part I

Theorizing Place and Mobiles

1 Mobilizing Place
Conceptual Currents and Controversies

Rowan Wilken and Gerard Goggin

"We are always in place, and place is always with us."[1]

"'Place' [. . .] in the radically imploded space of the global civilisation of the early twenty-first century remains [one] of the most problematical but compelling human concerns within the continuing experience of modernity."[2]

"Traditionally, place is an interweaving of communication and action."[3]

INTRODUCTION

Sometime in late 2010, the number of mobile cellular subscriptions worldwide exceeded the five billion mark, more than doubling since 2005. As the International Telecommunications Union has noted, mobile services are now available to over 90 percent of the world's population, including some 80 percent of the population living in rural areas.[4] Of course, such aggregate figures belie important aspects of actual access, consumption, and use of these technologies—especially of the myriad applications, functions, and meanings that comprise them. Nonetheless, there is abundant evidence of the global growth and ubiquity of mobile technologies.

From 2005 to the present day, we have also witnessed a fundamental change in the nature of mobile technologies themselves, made possible by the deployment of third-generation (3G) and fourth-generation (4G) networks, and the intelligence at the edges of such networks in the multimedia capabilities of the mobile technologies themselves. Since the arrival of the iPhone in mid-2007, and then the iPad in early 2010, the smartphone and tablet computer, respectively, have attracted the keen interest of users, industry insiders, and content developers alike. Such developments in the technical nature of mobiles have been intertwined with the shift from the mobile phone's innovations in voice and text communications to the central role mobiles now play in contemporary media—illustrated by the rise of many distinctive mobile-remediated forms: the resilient medium of text messaging, especially important in low-bandwidth countries in the global south; mobile news and citizen journalism; video and television; music and games; maps, location, navigation, and wayfindings; mobile Internet, in many different forms; apps for the care of the self, whether to do with life-style, identity, fitness, health and illness; and e-reading on mobile devices. In these developments in mobile technologies, place has played a prominent role. Because these are technologies in which mobility is key, there has been an everyday concern with their relationship to particular places (e.g., the space occupied by an individual while they use a device) and place in general. As mobile technologies have developed, their links with, focus on, and reliance on place have only deepened. This is most obvious in the wave of location, mapping, and sensor technologies that underpin applications such as maps for mobiles and Internet-based geoweb applications migrating to mobile platforms (most prominently, Google Maps).

There is intense popular, commercial, government, and civil society interest in mobile technologies. In research, mobile technologies have been examined extensively from a variety of perspectives across many disciplines. In particular, much has been revealed about the implications for mobile technologies on space and time.[5] Surprisingly, there has been a great deal less research and thinking on these technologies and the important role of place. Thus, the interactions between mobile technologies and place are little understood, despite the recent renewal of interest in related questions of location and geography in fields such as media and communications, cultural studies, and sociology.[6]

In this light, the book gives detailed, critical attention to the concept of place, exploring what significance this concept has in an era that sees continued and rapid growth in networked mobile technologies, and how these technologies have contributed to reconfigured understandings of and engagements with place. To explain the key theories and conceptual issues in mobile technologies and place, this chapter takes three progressive steps. First, the chapter begins by examining this most elusive of concepts—place—and how it has been conceived and debated over time by a wide range of theorists and critics from the fields of humanist geography (Yi-Fu Tuan, Edward Relph, Nigel Thrift, Doreen Massey, Tim Cresswell), architecture and phenomenology (David Seamon), philosophy (Edward Casey, Jeff Malpas), sociology (Manuel Castells), anthropology (Marc Augé), critical theory (Michel de Certeau), and media and communications (Joshua Meyrowitz). Second, we consider the particular challenges that are posed to established understandings of place by various forms of (macro and micro scale) mobilities *and* global networked information and communications technologies—in other words, the dual shaping influence of what Thomas Misa evocatively terms "the compelling tangle of modernity and technology."[7] Third, we narrow the focus slightly to consider the impacts of *mobile technologies* on place. The chapter concludes by returning to questions of definition, and, in overall terms, what is at stake in this concept of place in the present era of global mobile technologies. Let us begin, then, by developing an account of place: how it has been defined and understood, and how it forms a rich yet elusive concept.

UNDERSTANDING PLACE

The concept of place is notoriously complex and fraught. According to Jody Berland, it "has become one of the most anxiety-ridden concepts today."[8] Dealing with the idea of place is especially difficult due to its lack of definitional clarity and precision. As Nigel Thrift notes, "defining exactly what place is turns out to be very difficult."[9] Even in as compendious and impressive a study as Edward Casey's *Getting Back into Place*, place is everywhere present but nowhere defined.[10] Basic dictionary definitions do little to resolve general understanding of the term. For instance, *The Australian Concise Oxford Dictionary* offers thirteen variations, which range from broad references to space and its occupation to the differentiation of types or subcategories of geographical space and the occupation of these spaces (including, in order of increasing expansion: a residence or dwelling; a group of houses in a town; a town square; a village, a town, a city; an area or region).[11] This particular example adds support to Tim Cresswell's observation that "the commonsense uses of the word place belie its conceptual complexity."[12]

As Cresswell explains, in practical terms, place can be usefully understood as "a meaningful site that combines location, locale, and sense of place."[13] These last three elements correspond in turn with the following: "the 'where' of place"; the "material setting for social relations"; and, "the more nebulous meanings associated with a place: the feelings and emotions a place evokes."[14] Thus, as Edward Relph puts it, places "are constituted in our memories and affections through repeated encounters and complex associations."[15]

One of the issues with definitions of place, and with attempts at definition in general, as the preceding explanation highlights, is that place "is not just the 'where' of something; it is the location plus everything that occupies that location seen as an integrated and meaningful phenomenon."[16] Lukermann describes this as a process involving complex integrations of nature and culture that have developed and continue to develop in particular locations, and which connect flows of people and goods to other places.[17]

In other words, "in any given place we encounter a combination [and interlinking] of materiality, meaning and practice."[18] Each of these three terms warrants further explanation. The first, materiality, is intended to suggest that places have both "a material structure," as well as being constituted via "all the material things that pass through" them, such as "commodities [including mobile media technologies], vehicles, waste, and people."[19] The second, meaning, connects back to one of the most entrenched understandings of place from within human geography—the idea that "location became place when it became meaningful."[20] The third and final element, practice, and more specifically *mundane practice*, is especially significant for understanding the interactions of mobile technology and place.[21] As Cresswell writes:

> Places are continuously enacted as people go about their everyday lives—going to work, doing the shopping, spending leisure time, and hanging out on street corners. The sense we get of a place is heavily dependent on practice and, particularly, the reiteration of practice on a regular basis.[22]

It is this aspect of place that forms a vital ingredient in the examination of mobile technology and place in this book.

An even more expansive perspective on place is to suggest that difficulties of definition and experiential expression are in fact due to the notion of place being all-pervasive, structuring and shaping every facet of our lives, as well as of our negotiation and experience of the lived world.[23] In this respect, difficulties in grasping the notion of place are very much like the difficulties attending the category of the quotidian or everyday.[24] As Maurice Blanchot says of the everyday, "whatever its other aspects, the everyday has this essential trait: it allows no hold. It escapes."[25] Its pervasiveness renders it as platitude.[26] But, as Blanchot adds, "this banality is also what is most important, if it brings us back to existence in its very spontaneity and as it is lived."[27] And so it is with place. The pervasiveness of place and its plurality of forms means that it allows no hold; its ubiquity and diffuseness is what makes place most important as it informs and shapes lived existence.[28] Thus, place can be understood as all-pervasive in the way that it informs and shapes everyday lived experience—including how it is filtered and experienced via the use of mobile media technologies. As Casey puts it, "place serves as the condition for all existing things."[29]

Further valuable conceptualizations of place include: the idea of place as being a structure of feeling or generated by "fields of care;"[30] place as open and relational rather than bounded;[31] place as "more and more demonstrably the outcome of embodied social practice," where "people determine its shape and meanings;"[32] and, inverting the previous understanding, as that which "is integral to the very structure and possibility of experience"[33] and "within and with respect to which subjectivity is itself established."[34]

These quite diverse definitional tributaries point to the difficulty in arriving at shared definitional understandings of place. This task is made all the more difficult, Thrift argues, by two further factors that are increasingly hard to disentangle: the fact that meanings of place are "increasingly bound up with the growth of the media," including "the burgeoning of representations of place,"[35] and the fact that place experience has been inexorably altered as a result of both micro and macro forms of mobility. That is to say, place has been affected profoundly by the micro scale of (largely localized) experiences of networked mobility (the primary concern of a number of the contributions to this book), and the macro scale of global geopolitical transformations;[36] as well as, the micro politics of mobile, technologically equipped bodies in transit through place(s),[37] and the macro-scale geopolitics of voluntary and forced migration and displacement.[38] In the following sections, we explore some of this terrain, beginning with a consideration of the impacts of larger scale mobilities on place as understood within the established literature on globalization, before turning to the more intimately scaled impacts of mobile technology use on place.

GLOBAL FLOWS: MOBILITY, TECHNOLOGY, PLACE

Place and mobility, Tim Cresswell observes, tend to be "marked by disagreements between those who see mobility and process as antagonistic to place and those who think of place as created by both internal and external mobilities and processes."[39] Expanding on this observation, Massey and Thrift suggest:

> place has become one of the key means by which the social sciences and humanities are attempting to lever open old ways of proceeding and tell new stories about the world, whether as a vast space of flows [more on this to follow] in which place is gradually being erased or as the sensuous rediscovery of the pleasures of the specific in which place is being rediscovered as something in-between.[40]

Both these passages draw out how (macro scale) *mobility*—especially of the kind associated with globalization—forms a kind of fault line along which place is, depending on one's point of view, either unsettled or reconstituted.

The potentially detrimental effects of global processes on place were a major concern for key scholars of place, such as Edward Relph and David Seamon, who were writing in the mid to late 1970s, an influential period in North American scholarship on place. Perhaps influenced by the contemporaneous wave of equally influential media and communications scholarship on cultural imperialism,[41] which expressed the idea that cultural globalization "can act as a solvent, dissolving cultural differences to create homogeneity across the globe,"[42] Relph and Seamon both express concern for a potential gradual erasure of place and the specificity and uniqueness of place. For instance, Seamon writes in his 1979 book, *A Geography of the Lifeworld* (a book that was strongly influenced by the arguments of Relph's *Place and Placelessness*, published in 1976), that "technology and mass culture destroy the uniqueness of places and promote global homogenization."[43] This has led to the charge that Seamon is "suspicious of developments in 'mass communications',"[44] an accusation that he resists.[45] Even so, Seamon has softened his position more recently, arguing his larger point is really that "rapid societal and technological changes" permit people to "become free of the habitual embodiment to place" that, prior to modern networked communications, "had always been an integral part of human life everywhere."[46] This has proven a very compelling argument, and one that continues to find purchase. For instance, this perspective is developed in the influential work of Joshua Meyrowitz (to be discussed later in this chapter), and is also meditated on by Edward S. Casey in Chapter 10, this volume.

Turning to the available critical literature on cultural globalization from the 1990s (in many ways the high point of globalization scholarship), here place manifests itself and is understood in complex and often contradictory ways. This is illustrated most clearly (and most famously) in the work of Manuel Castells, where he juxtaposes what he terms the "space of places" with the "space of flows." Castells defines the "space of places" as the "historically rooted spatial organization of our common experience" whereas the "space of flows" is understood as a series of transformations involving the circulation of capital, information, technology, organizational interaction, and images, sounds, and symbols.[47] Flows, he then argues, "are not just one element of the social organization: they are the expression of processes *dominating* our economic, political, and symbolic life."[48]

However, to describe these two phrases as "juxtaposed" understates the extent to which there is significant interaction, overlap, and movement between them. For instance, to cite one example, Castells writes that, on the one hand:

> Localities become disembodied from their cultural, historical, geographic meaning, and reintegrated into functional networks, or into image collages, inducing a space of flows that substitutes for the space of places.[49]

On the other hand, at a later point, Castells qualifies this perspective, writing:

> The space of flows does not permeate down to the whole realm of human experience in the network society. [. . .] The overwhelming majority of people, in advanced and traditional societies alike, live in places, and so they perceive their space as place-based. A place is a locale whose form, function and meaning are self-contained within the boundaries of physical contiguity.[50]

What these tensions highlight is that "the new industrial system is neither global nor local" but "a new articulation of global and local dynamics."[51] It is in this context that "glocalization" has emerged as an important notion for capturing global-local tensions.[52] The term was imported into social and cultural theory by Roland Robertson to correct what he saw as one of the "major weaknesses" attending use of the word "globalization:" that is, "the tendency to cast the idea of globalization as inevitably in tension with the idea of localization."[53] This is problematic, he argues, as it denies the full extent to which "the concept of globalization has involved the simultaneity and the interpenetration of what are conventionally called the global and the local."[54] Thus, for Robertson, the term "glocalization" serves a specific, rhetorical purpose: it works to emphasize the point that the original concept of globalization has involved and will continue to involve "the creation and the incorporation of locality, processes which themselves largely shape, in turn, the compression of the world as a whole."[55] What is clear from Robertson's arguments is that to engage with globalization and the "space of flows," to use Castells's specific phrase, requires an ongoing concern for singularities and particularities of place and locality, and for "the ways in which 'the local and the global are always interlocked and complicitous.'"[56] In short, to extrapolate this to the notion of place, while in many respects a troublesome and contested term, it is nonetheless indispensable,[57] and, to adapt the words of Casey,[58] we require places in which to exist; we are immersed in places and could not do without them.

In this sense, cultural globalization debates are striking for the way that they have led to a renewed concern for place. This revival of interest in the concept of place—which parallels a recent "geographical turn" in media studies[59]—has direct implications for thinking about how we engage with mobile media. As Larissa Hjorth writes, "the dynamic interaction between globalization and practices of locality is nowhere more apparent than in debates surrounding mobile telephony and its dissemination and appropriation at the level of the local."[60] For her part, Susan Friedman suggests that the "local and the global are always already interlocking and complicitous."[61] The deep engagement of geography, especially cultural geography, with questions of place, is something that Chris Gibson, Susan Luckman, and Chris Brennan-Horley emphasize in Chapter 7, this volume. In their contribution, which discusses telecommunications in the tropical north Australian city of Darwin, they argue that "theorization about how mobile technologies might transform places itself needs to put considerations of geography at the centre of things—rather than at the end of a simplistic chain of causality that link people to technology, to (reconfigured) place."[62]

This is also very much the perspective evidenced within British geographical scholarship. Reflecting back on the global transformations of the 1980s and 1990s, Massey and Thrift describe how these conditions led to the formulation of a revised, "relational" understanding of place:

[A]n imagination of space as constructed out of nets of interconnections began to take shape. This had major implications for the conceptualization of place. Place could no longer be taken as a given, as an already bounded entity whose characteristics were somehow produced within it (out of the soil, as it were). Rather a "place" was now, in this view, understood as being a moment in a wider relational space; its very specificity was constructed out of relations with elsewhere [. . .]; the character of a place was always in construction.[63]

From this perspective, place can be understood as a bounded but open and contested site, a complex product of competing discourses, ever-shifting social relations, and internal (as well as external) events; Massey elsewhere refers to this as the "throwntogetherness" of place, the "event" of place.[64] In other words, any given place is "dependent upon the interconnected-ness of the elements within it—as it is also dependent on its interconnec-tion with other places."[65] What is articulated here, then, is a formulation of place that is imagined as "articulated moments in networks of social relations and understandings," albeit which carry implications far beyond "what we happen to define for that moment as place itself."[66]

The possibilities presented by a relational understanding of place for media and communications research has been recognized by a number of scholars.[67] In the present context, relational place has considerable potential for thinking about the cultural politics of various forms of everyday media use within the broader context of globalization. It also has specific value for understanding the use of mobile devices, such as the cell phone, which can be grasped by referencing Mizuko Ito's research on Japanese youth and their mobile use. This research revealed phone use that involves a com-plex set of interactions within and between places, incorporating physical transportation (on foot and by other means), navigation (utilizing, among other tools, existing physical landmarks), fluid temporal arrangements, and communication (both mediated and face-to-face).[68] All of these maneuvers implicate particular places (the ultimate destination of the mobile caller, as well as those through which they pass) in much wider geographical and informational networks, as well as networks of social relations and under-standing. The concept of networks in relational thinking of place, as well as other topologies (such as region, network, fluid, and fire), is the subject of Richard Ek's "Topologies of Human-Mobile Assemblages," Chapter 3, this volume.

As we begin to move from macro-scale debates on place and globaliza-tion toward more intimately scaled engagements with mobiles and place, there are two intermediary sets of ideas that are important to consider: the first involves a brief account of (and response to) Michel de Certeau's unique conception of space and place; the second concerns Marc Augé's influential (if increasingly contested) notion of "non-places."

According to Tuan's influential (if problematic) definition of place, "what begins as undifferentiated space becomes place as we get to know it better and endow it with value."[69] For Michel de Certeau, something like the opposite is true. In his seminal text, *The Practice of Everyday Life*, de Certeau defines place as a "'proper' and distinct location," "an instantaneous configuration of positions" which "implies an indication of stability."[70] Space, meanwhile, is said to exist "when one takes into consideration vectors of direction, velocities, and time variables" and "intersections of mobile elements."[71] Thus, for de Certeau, "in short, *space is a practiced place*."[72] Such a formulation of space as "practiced place" is understandable in the context of his study of the multiform richness associated with pedestrian movement, and is worked through in complex ways.[73] Indeed, it has been noted that, when engaging with this work, one ought to acknowledge how de Certeau pursues a "sustained orchestration of binary terms" that works to "challenge the structures of binary thought."[74] Even with this in mind, however, the difficulty with his recasting of the place/space relation is that it is premised on a very conservative conception of place which, as we have seen in relation to geographical thought, is arguably no longer tenable. As Thrift argues, "increasingly, how we think of place is connected to mobility" at a variety of scales.[75] Thus, in relation to pedestrian mobility and mobile media use, the long-standing conception of *stabilitas loci* may itself need to be reformulated as *mobilitas loci*: "the difference between place experienced as stable (if not fixed), to multiple places experienced in and through mobility."[76] That is to say, de Certeau arguably underestimates the "nested" qualities of place and its relational dimensions, constitution, and continual reconstitution. Moreover, as Caroline Bassett argues, "it is clear that the (negative) place accorded to information technologies in de Certeau's consideration on the dialectics of power, control and freedom is challenged by the case of the mobile."[77]

Meanwhile, for the French anthropologist Marc Augé, many of the locations that have been opened up to support and service global mobility and modernity should not, strictly speaking, be considered as "places." Rather, they ought to be more accurately characterized, in his terms, as "non-places." According to Augé, the contemporary cultural landscape of globalization is characterized by an overabundance of information and a growing tangle of interdependencies which leads to the creation of an "excess of space correlative with the shrinking of the planet."[78] Augé coins the term "non-places" to describe this expanding excess. "Non-places" are those interstitial zones where we spend an ever-increasing proportion of our lives: in supermarkets, airports, hotels, cars, on motorways, and in front of ATMs, TVs, and computers. For Augé, such "non-places" are the real measure of our time. The true extent of them can be quantified, he writes,

> by totalling all the air, rail and motorway routes, the mobile cabins called "means of transport" (aircraft, trains and road vehicles), the airports and railway stations, hotel chains, leisure parks, large retail outlets, and finally the complex skein of cable and wireless networks that mobilize extraterrestrial space for the purposes of a communication so

peculiar that it often puts the individual in contact only with another image of himself.[79]

As Poster explains, such spaces, according to Augé, "permit individuals to have presence only by dint of passports, credit cards, travel tickets and the like, undermining the human attachment to location."[80] Thus, these "non-places," in Thrift's words, are viewed very critically by Augé as "empty, uniform, solitary: multiple, certainly, but pointlessly so."[81]

There are a number of difficulties with this formulation of "non-places." For instance, as Spinney points out, firstly, while a great number of people engage with these locations in a transitory, mobile way, "they are real and people live their lives in and through them, increasingly so in ever more mobile societies."[82] Secondly, Augé also seems to underestimate human adaptability—the way, for instance, we carve out places within the everyday sites of transit and mobility[83]—and how, even in transit, people can form certain kinds of place attachments to liminal locations (like airports and supermarkets), and, additionally, how these same liminal sites are also often tied in complex ways to the personal and professional identities and techno-social interactions[84] of those passing through them. Thirdly, in this model, "movement itself seems to have been largely ignored as a social practice generative of meaning in itself."[85] Spinney's contention is that meaning is created "through an embodied and sensory engagement with place which does not rely solely on notions of landscape, dwelling or sociality."[86] Rather, this meaning making occurs, he argues, "moment by moment through a series of fleeting and solitary embodied encounters."[87] Accordingly, how we experience the "character of a place is dependent upon *how* we are in a place, and how we perceive and organise sensory input," not just "between bodies and environments" but also how "they are mediated by other everyday entities and technologies,"[88] including—indeed, *especially*—mobile technologies.

In a related vein, Varnelis and Friedberg[89] take issue with Augé's assessment that "non-place is conducive to 'solitary' as opposed to the traditional 'collective' contractual obligations based on shared values and beliefs."[90] In the present digital age, they write, his idea of solitary non-places "are an artifact of the past."[91] Things, they suggest, are more complicated. Aided by mobile media and other networked information and communication technologies, the sorts of locations that Augé describes as "non-places" can be both material and "virtual," experienced as either solitary or as part of a wider (but not necessarily physically co-present) social network (more on this later).[92] For Thrift, such insights are suggestive of what he calls "the other side of the decomposition of place:" that is to say, that "the world's places are not steadily growing more inauthentic," or counterposed by a proliferation of "non-places,"[93] but rather that the places we encounter might be understood as "recombining in new ways."[94] Consideration of mobile technologies and their uses gives insight into how this recombination is or might be occurring.

TEXTURES OF PLACE: THE (MICRO) PRACTICES OF MOBILE MEDIA USE

As we mentioned earlier, place is an obvious feature of mobile communication, most obviously because communication is occurring in different locales than it did previously, when telephony occurred in clear relation—signified by the telephone cord—to fixed-line equipment and infrastructure. The places of making and receiving a phone call, or indeed of data communications or videotelephony, were relatively well-defined: the payphone booth or cabinet; the office desk; the hall or bedside table; or the much rarer transmitter allowing trunked radio communications in a car, vehicle, or boat. With the arrival of cellular mobile telephony from the late 1970s, there was a need for social innovation to occur to accompany the technical invention (a process that began from at least the nineteenth century with telegraph and wireless, as well as the telephone). The immediate places—small contexts—of telephony proliferated with the advent of the mobile, compelling users and their societies to reconstruct the relations with place and mobile, and indeed to understand the kind of shaping—"tuning" (to borrow Richard Coyne's term)[95]—of place unfolding in concert with the technology.

Scholarship on mobile phones and mobile media has, over the past decade or so, flowered and matured. Much of this research addresses—though perhaps more implicitly than explicitly or systematically—issues pertaining to place and place experience.[96] With the crisscrossing of mobiles and the Internet, there is now also an emerging body of work that attends to the new issues concerning place that this poses—especially in relation to location technologies.[97] Of this research, here we wish to limit our focus to two strands which, for the purposes of the present discussion, can be grouped under the rubrics of phenomenological accounts of body-technology, or socio-technical, relations on the one hand, and questions of presence on the other.

Phenomenological Accounts of Body-Technology Relations

In a 2003 essay, Joshua Meyrowitz writes: "With computers, mobile phones, and the internet, many different tasks [. . .] take place in no particular place, and involve the same basic position and movements of body, head, and hands."[98] Almost as if in direct response to Meyrowitz, three years later, in a 2006 article reflecting back on his earlier book, *A Geography of the Lifeworld*, phenomenologist David Seamon makes the following plea:

> One need is thorough phenomenologies of the kinds of encounter that various media and digital communications involve. Clearly, different media, all other things being equal, change the kind and intensity of encounter that the listener, watcher, or user has with the 'real' world in which the listening, watching, or use takes place.[99]

It is precisely these kinds of questions that have formed the basis of the work of critics such as new media scholar Ingrid Richardson. In her work, Richardson explores:

> A number of phenomenological issues relating to second- and third-generation mobiles, including their unique screen properties, the emergent screen-body ontology specific to mobile phone media, their capacity to reconfigure our relation to actual and virtual space, and their particular impact on our experience as-bodies.[100]

Also crucial to these phenomenological engagements with mobiles and place is the question of how to account for fluctuating levels of attentiveness. Margaret Morse has referred to this as "an ontology of everyday distraction"[101]—the idea that a precondition of our engagement with technology, space, and place is an experience of distraction and disjointedness. It is an understanding that is connected with, and qualified by, what Caroline Bassett designates as "questions of attention"[102]—that is, how our level of attention with our technological tools and our surroundings is influenced strongly by a variety of environmental, contextual, interpersonal, and cognitive factors. Thus, for technologically equipped pedestrians, for instance, attention is always a fluctuating state, where pedestrians tend to switch their attention from one place or activity to another, each time at the expense of the last and as a result of competing stimuli. On one hand, this can mean that mobile users pay more attention to "phone space," which is "often prioritized over local space."[103] On the other hand, however, "since attention never presumes absolute presence, it cannot presume absolute disconnection," rather there is a toggling between the two.[104] This is evident in various ethnographic studies of mobile phone use in public space that report numerous instances of such shifting between sights, sounds, tasks, and other inputs. In Höflich's study of mobile phone users in Italy's Piazza Matteotti, he brings the two perspectives together in his observations of "quasi-autistic behavioural patterns, [of people] who—holding a mobile phone to their ear—do not want to know anything of the world around them" while also acknowledging "a certain 'sense of place'" structures their users' mobile-mediated engagements with the Piazza.[105]

Everyday distraction is also closely tied to the specific and complex nature of user engagement with mobile handheld screen devices. To give one example, a particularly useful model for making sense of the specificity of our visual engagement with mobile screens is Heidi Rae Cooley's notion of "screenic seeing."[106] This is the idea that vision is "no longer a property of the window and its frame," but, rather, contributes to a more expansive "material experience of vision" where "hands, eyes, screen, and surroundings interact and blend in syncopated fashion."[107] Marrying this with Walter Benjamin's notion of tactility, where the tactile is associated with distraction and habit, Cooley argues that "screenic seeing" is "never focused." Rather, it "spreads out, across various images (of landscape and screen), as a result of and in tandem with the interpenetration of the hand and MSD [mobile screenic device]."[108] In short, the complexities of these interactions suggest that, mobile devices, such as the cell phone, involve their own forms of "body-technology relations"[109] that require accounting for, as do their impacts on our apprehension and experience of presence (both face-to-face and remote—the focus of what is to follow), space and place, the relations between bodies, and human agency.[110] Richardson joins with Rowan Wilken to draw on post-phenomenological thinking to argue that mobile media devices "call for a distinctly different set of body-screen-place relations than those of many other forms of (especially fixed or largely stationary) screen-based media"[111] in Chapter 11, this volume. In doing so, they persuasively show that "a sense of place—if somewhat altered—remains integral to the body-mobile relation, as it is both 'contained' in the micro worldliness of the device and hybridized by the perceptual merger of situated 'hereness' with online networks."[112]

Furthermore, what Seamon is calling for in his earlier passage, and what we see already evidenced in the research noted previously, is the kind of work that considers in detail what he has called "body routine"[113] (i.e., "a set of integrated gestures, behaviors, and actions that sustain a particular task or aim"[114]), which contributes to what he calls a "place ballet"[115] (i.e., "an interaction of time-space routines and body routines rooted in space, which becomes an important place of interpersonal and communal exchanges, actions, and meanings"[116]). Such rhythms as they pertain to mobile media—evoked by James Katz's phrase "the choreography of mobile communication"[117]—are the subject of Iain Sutherland's study of friends settling in Barcelona, "Connection and Inspiration: Phenomenology, Mobile Communications, Place" (Chapter 9, this volume). For Sutherland, it is "through ubiquity and entry into the domain of the everyday, [that] mobile technologies may become enfolded into pre-conscious regimes of interaction becoming part and parcel of a bodily perceptual apparatus through which place is experienced."[118]

The second strand of mobile technology research we want to consider briefly here concerns questions of presence.

Presence

Presence has long been considered an important concept for information and communications technology researchers, especially those with an interest in computer-mediated communications and its antecedents.[119] In research on mobile technology, presence has also emerged as a crucial concern. This interest is manifest in a variety of ways: in relation to concerns over the privileging of the individual over the collective, or what Gergen calls "absent presence;"[120] in relation to how "presence" is negotiated via short message service (SMS) use, or what Hjorth terms "postal presence;"[121] in relation to the "always on"[122] nature of mobile phone use, or what Licoppe terms "connected presence;"[123] in relation to observed interactions between those who are physically co-present and those who are located elsewhere but are nonetheless "wirelessly co-present;"[124] and, finally, in relation to at times delicately balanced negotiations between phone use and situational context.[125]

A key example of the last idea listed here is the question many mobile phone calls begin with: "Where are you?" This particular question is important in discussions of place and mobiles insofar as it points to the complex and quite subtle ways that everyday conversation has long been understood to be in crucial ways place-focused—something that was explored in Emanuel Schegloff's pioneering study of "locational formulations" in utterances.[126] Eric Laurier has drawn on Schegloff's work to argue that one of the key reasons for the "Where are you?" question in relation to mobile use is to establish "some sense of shared context."[127] According to Laurier, for people who know each other, temporal orderings are generally already established, "what remains . . . are their spatial relations."[128] In other words, a "spatiotemporal context is being mutually accomplished"[129] during the course of the conversation and, for this reason, Laurier notes, "knowing at the outset whether one is calling a mobile or not" is important insofar as it "shapes the character of ordinary geographical work that will need to be done by caller and by called."[130]

According to Laurier, the fact that this "ordinary geographical work" is necessary is due to the fact that "what mobile phones bring is a loss of the tie between place and [person . . .] as the caller is likely not to be calling a place but a person."[131] The suggestion here is that mobile phones operate independent of place [132] and thus mark a shift from place-to-place to person-to-person communication.[133] What is neglected in this formulation is a full appreciation of what is carried by the fact that "communication always *takes place* somewhere, in particular social and spatial contexts, and place is always 'in' the communication."[134] In other words, we might say that the "geographical work" being done between caller and called is in fact far from "ordinary," and what sits behind the "Where are you?" question (at least in part) is an implicit understanding that the "place" of call reception matters a great deal.[135]

Meyrowitz uses the same "Where are you?" question to make a related but slightly different point to that of Laurier: to highlight the complexities of place and presence in the age of mobiles and how "we are, in a way, both inside and outside the locale at the same moment."[136] Building on his earlier arguments in *No Sense of Place*, where he writes that electronic media "lead to a nearly total dissociation of physical place and social 'place,'"[137] in a more recent essay Meyrowitz echoes the sentiments of David Seamon in arguing that, "the more our sense of self and experience is linked to interactions through media, the more our physical locales become the backdrops for these other experiences rather than our full life space."[138]

This may well be so. As Varnelis and Friedberg observe, "place, it seems, is far from a source of stability in our lives" and, "once again, is in a process of a deep and contested transformation."[139] Nevertheless, this fact, they argue, does not diminish the significance of place. Rather, "place is as important as ever, playing a key role"[140] in our interactions with information and communications networks. Indeed, for some critics, the fact that we are (in Meyrowitz's words) "both inside and outside the locale at the same moment" ought to prompt us to rethink our engagements with place as "hybridized" and which "merge the physical and the digital in a social environment created by the mobility of users connected via mobile technology devices."[141] Thus, as Gordon and de Souza e Silva argue, "location-aware technologies allow us to seamlessly interact with the remote and the proximate" in ways that complicate attempts to draw clear delineations between "co-presence" and "networked interaction" (Chapter 5, this volume).[142] According to Paul Dourish, these developments do not create new places, but rather "allow people to encounter and appropriate existing [places] in different ways" that "transform them" as "sites of everyday action."[143] This makes for a complex "encoding of place" something which, as Gerard Goggin argues (Chapter 12, this volume), is a prompt for reconsidering the histories and affordances of mobile technologies, and their role in the contemporary media ecologies of location being created by the intersection of mobiles and the Internet. For Francesco Lapenta, in his striking formulation (Chapter 13, this volume), these "location-based media or geomedia are to space what the watch is to time"—powerful, normative renegotiations of bodies and their constitutive relations to place.[144]

CONCLUSION: MOBILIZING PLACE

This chapter forms a backdrop against which the overall aims of this book can be understood. This collection seeks to examine the interrelationship between two key phenomena: the notion of and our experiences of place on the one hand, and contemporary mobile media use on the other hand. The chapters that follow, then, provide a comprehensive examination of the complex interactions between mobile media technologies and issues of place. The collection aims to develop a rigorous and detailed understanding of how theories and experiences of place interact with practices of mobile technology use, and to this end the work of key place theorists is set in critical dialogue with the work of leading mobile technology scholars. The purpose of this concertedly dialogic approach is to develop a fuller, critical understanding of the intersections and interconnections between mobile technology use and notions of place to understand what significance both elements have for understanding and living in the digital age.

A central premise of this book is that both of these phenomena are hugely significant in their own right—place is considered fundamental to the construction of our life histories and what it means to be human, while mobiles now form an intrinsic part of the daily lives and habits of billions of people worldwide—*and* for the manifold ways that they mutually inform and shape each other. Place is a notion that is of enduring relevance—one worth mobilizing—if we are to comprehend fully how we think about and experience who we are, where we are, and the ways we interact and relate with one another. In other words, place is considered a vital notion in that it represents a "weaving together" of social and human-environment interactions in ways that situate place as central to how embodied, technologically mediated mobile social practice is understood. The scholarship gathered together in this collection, then, it is hoped, will make an important contribution to knowledge on the interactions between mobile technologies, their uses, and place; will challenge the way place has historically been understood to operate in relation to the social as well as wider uses of mobile technologies; and, finally, will work to underscore what we see as the enduring importance of ideas of place in the present age, along with the urgent need to engage critically with these ideas.

NOTES

1. Joshua Meyrowitz, "The Rise of Glocality: New Senses of Place and Identity in the Global Village," in *A Sense of Place: The Global and the Local in Mobile Communication*, ed. Kristóf Nyíri (Vienna: Passagen Verlag, 2005), 21.
2. Peter Scriver, "On Place," in *De-Placing Difference: Architecture, Culture and Imaginative Geography*, ed. Samer Akkach (The University of Adelaide: Centre for Asian and Middle Eastern Architecture, 2002), 4.
3. Paul C. Adams, "Network Topologies and Virtual Place," *Annals of the Association of American Geographers*, 88(1) (1998), 94.
4. International Telecommunications Union, "The World in 2010: ICT Facts and Figures," http://www.itu.int/ITU-D/ict/statistics/.
5. For a fine set of papers on space and time in mobile communications, see Rich Ling and Scott Campbell, eds., *The Reconstruction of Space and Time: Mobile Communication Practices* (Piscataway, NJ: Transaction, 2009).
6. See, for example: Paul C. Adams, *Geographies of Media and Communication* (New York: Wiley-Blackwell, 2009); Brett Christophers, *Envisioning Media Power: On Capital and Geographies of Television* (Lanham, MD: Lexington, 2010); Richard Coyne, *The Tuning of Place: Sociable Spaces and Pervasive Digital Media* (Cambridge, MA: MIT Press, 2010); Nick Couldry and Anna McCarthy, eds., *MediaSpace: Place, Scale and Culture in a Media Age* (New York: Routledge, 2004); Greg Elmer, Charles H. Davis, Janine Marchessault, and John McCullough, eds., *Locating Migrating Media* (Lanham, MD: Lexington, 2010); Jesper Falkheimer and André Jansson, eds.,

Geographies of Communication: The Spatial Turn in Media Studies (Götenberg: Nordicom, 2006); David Morley, *Home Territories: Media, Mobility and Identity* (New York: Routledge, 2000); David Morley and Kevin Robins, *Spaces of Identity: Global Media, Electronic Landscapes, and Cultural Boundaries* (New York: Routledge, 2005); Hamid Naficy, ed., *Home, Exile, Homeland: Film, Media, and the Politics of Place* (New York: Routledge, 1999); Anna Roosvall and Inka Salovaara-Moring, eds., *Communicating the Nation: National Topographies of Global Media Landscapes* (Götenberg: Nordicom, 2010).

7. Thomas Misa, "The Compelling Tangle of Modernity and Technology," in *Modernity and Technology*, eds. Thomas Misa, Philip Brey, and Andrew Feenberg (Cambridge, MA: MIT Press, 2003), 1–30.

8. Jody Berland, "Place," in *New Keywords: A Revised Vocabulary of Culture and Society*, eds. Tony Bennett, Lawrence Grossberg, and Meaghan Morris (Oxford: Blackwell, 2005), 257.

9. Nigel Thrift, "'Us' and 'Them': Re-imagining Places, Re-imagining Identities," in *Consumption and Everyday Life*, ed. Hugh Mackay (London: Sage, 1997), 196.

10. Edward S. Casey, *Getting Back into Place: Towards a Renewed Understanding of the Place-World* (Bloomington, IN: Indiana University Press, 1993).

11. J. M. Hughes, P. A. Michell, and W. S. Ramson, eds, *The Australian Concise Oxford Dictionary* (Melbourne: Oxford University Press, 1992), 863; for more detailed discussion, see Edward S. Casey, *The Fate of Place: A Philosophical History* (Berkeley, CA: University of California Press, 1997).

12. Tim Creswell, "Place" (2009), http://www.elsevierdirect.com/brochures/hugy/SampleContent/Place.pdf

13. Cresswell, "Place," 1.

14. Cresswell, "Place," 1.

15. Edward Relph, "Geographical Experiences and Being-in-the-World: The Phenomenological Origins of Geography," in *Dwelling Place and Environment: Towards a Phenomenology of Person and World*, eds. David Seamon and Robert Mugerauer (Dordrecht: Martinus Nijhoff Publishers, 1985), 15–32; for an extended personal reflection on this understanding of place, see: bell hooks, *Belonging: A Culture of Place* (New York: Routledge, 2009).

16. Edward Relph, *Place and Placelessness*, reprint and third imprint (London: Pion, 1986), 3; see also, Christian Norberg-Schulz, "The Phenomenon of Place," *Architectural Association Quarterly*, 8(4) (1976), 3–10.

17. Fred E. Lukermann, "Geography as a Formal Intellectual Discipline and the Way in Which It Contributes to Human Knowledge," *Canadian Geographer*, 8(4) (1964), 167–172.

18. Cresswell, "Place," 1.

19. Cresswell, "Place," 1.

20. Cresswell, "Place," 1.

21. Cresswell, "Place," 2.

22. Cresswell, "Place," 2.

23. Casey, *Getting Back into Place*.

24. Maurice Blanchot, "Everyday Speech," trans. Susan Hanson, *Yale French Studies*, 73 (1987): 14.

25. Blanchot, "Everyday Speech," 14.

26. Blanchot, "Everyday Speech," 13.

27. Blanchot, "Everyday Speech," 13.

28. And this is perhaps why Casey offers no concise definition of place. His suggestion seems to be that we reach an understanding of place only by taking a circuitous route: by studying "the perplexing phenomenon of displacement,

rampant throughout human history and especially evident at the present historical moment, only in relation to an abiding implacement" (1993: xiv).

29. Edward S. Casey, *Getting Back into Place*, 15.
30. Yi-Fu Tuan, *Space and Place: The Perspective of Experience* (Minneapolis, MN: University of Minnesota Press, 1977).
31. Doreen Massey, *For Space* (London: Sage, 2005).
32. Berland, "Place," 258.
33. Jeff Malpas, *Place and Experience: A Philosophical Topography* (Cambridge, UK: Cambridge University Press, 1999), 32.
34. Malpas, *Place and Experience*, 35.
35. Thrift, "'Us' and 'Them',", 160.
36. Tim Cresswell, *On the Move: Mobility in the Modern Western World* (London: Routledge, 2006); David Morley, "What's 'Home' Got to Do with It?: Contradictory Dynamics in the Domestication of Technology and the Dislocation of Domesticity," *European Journal of Cultural Studies*, 6(4) (2003), 435–458; John Urry, "Mobility and Proximity," *Sociology*, 36(2) (2002): 255–274.
37. For detailed discussion, see Aharon Kellerman, *Personal Mobilities* (London: Routledge, 2006), 128–144.
38. David Morley, *Home Territories: Media, Mobility and Identity* (London: Routledge, 2000); David Morley and Kevin Robins (eds), *Spaces of Identity: Global Media, Electronic Landscapes and Cultural Boundaries* (London: Routledge, 1995); John Urry, "Mobile Sociology", *British Journal of Sociology*, 51(1) (2000): 185–203; John Urry, *Sociology Beyond Societies: Mobilities for the Twenty-First Century* (London: Routledge, 2000).
39. Cresswell, "Place," 9.
40. Doreen Massey and Nigel Thrift, "The Passion of Place," in *A Century of British Geography*, eds. Ronald Johnston and Michael Williams (Oxford: Oxford University Press, 2003), 276–277.
41. Jeremy Tunstall, *The Media are American* (New York: Columbia University Press, 1977); Ariel Dorfman and Armand Mattelart, *How to Read Donald Duck: Imperialist Ideology in the Disney Comic* (New York: International General, 1975); Herbert Schiller, *Mass Communications and American Empire* (New York: Augustus M. Keeley Publishers, 1969); for discussion, see John Tomlinson, *Cultural Imperialism* (London: Continuum, 1991).
42. Lauren Movius, "Cultural Globalisation and Challenges to Traditional Communication Theories," *PLATFORM: Journal of Media and Communication*, 2(1), January (2010), http://journals.culture-communication.unimelb. edu.au/platform/resources/includes/vol2_1/PlatformVol2Issue1_Movius. pdf.
43. David Seamon, *A Geography of the Lifeworld* (New York: St. Martin's Press, 1979), 91.
44. Shaun Moores, "Media Uses and Everyday Environmental Experiences: A Positive Critique of Phenomenological Geography," *Particip@tions*, 3(2) (2006), http://www.participations.org/volume%203/issue%202%20-%20 special/3_02_moores.htm
45. David Seamon, "*A Geography of Lifeworld* in Retrospect: A Response to Shaun Moores," *Particip@tions*, 3(2) (2006) http://www.participations.org/ volume%203/issue%202%20-%20special/3_02_seamon.htm
46. Seamon, "*A Geography of Lifeworld* in Retrospect."
47. Manuel Castells, *The Information Age: Economy, Society and Culture: Volume 1: The Rise of the Networked Society* (Oxford: Blackwell, 1996), 412.
48. Castells, *The Information Age*, 412. What is worth noting here, as Rantanen points out, is that "there is practically no globalization without media and

communications" (Terhi Rantanen, *The Media and Globalization* [London: Sage, 2005], 4).

49. Castells, *The Information Age*, 375.
50. Castells, *The Information Age*, 423.
51. Quoted in Castells, *The Information Age*, 392.
52. Roland Robertson, "Glocalization: Time-Space and Homogeneity-Heterogeneity," in *Global Modernities*, eds. Mike Featherstone, Scott Lash and Roland Robertson (London: Sage, 1995), 25–44.
53. Robertson, "Glocalization," 40.
54. Robertson, "Glocalization," 30.
55. Robertson, "Glocalization," 40.
56. Roland Robertson, "The Conceptual Promise of Glocalization: Commonality and Diversity," *ART-e-FACT*, 4 (2005), http://artefact.mi2.hr/_a04/lang_en/theory_robertson_en.htm (accessed April 1, 2011).
57. Scriver, "On Place," 4.
58. Casey, *Getting Back into Place*, ix.
59. André Jansson and Jesper Falkheimer, "Towards a Geography of Communication," in *Geographies of Communication*, 9–25.
60. Larissa Hjorth, "Society of the Phoneur," *antiTHESIS* 15 (2005), 208.
61. Susan S. Friedman, *Mappings: Feminism and the Cultural Geography of Encounters* (Princeton, NJ: Princeton University Press, 1998), 110.
62. Chris Gibson, Susan Luckman, and Chris Brennan-Horley, "(Putting) Mobile Technologies in Their Place: A Geographical Perspective" (Chapter 7, this volume).
63. Massey and Thrift, "The Passion of Place," 281–282.
64. Doreen Massey, *For Space* (London: Sage, 2005), 181.
65. Malpas, *Place and Experience*, 39.
66. Doreen Massey, *Space, Place, and Gender* (Cambridge, UK: Polity Press, 1994), 154.
67. Rowan Wilken, *Teletechnologies, Place, and Community* (New York: Routledge, 2011); Rowan Wilken, "Mobilizing Place: Mobile Media, Peripatetics, and the Renegotiation of Urban Places," *Journal of Urban Technology*, 15(3), 39–55; Richard Ek, "Media Studies, Geographical Imaginations and Relational Space," in *Geographies of Communication*, 45–66; Jayne Rodgers, "Doreen Massey: Space, Relations, Communications," *Information, Communication & Society*, 7(2) (2004), 273–291.
68. Mizuko Ito, "Mobile Phones, Japanese Youth, and the Re-Placement of Social Contact," in *Mobile Communications: Renegotiation of the Social Sphere*, eds. Rich Ling and Per E. Pedersen (London: Springer, 2005), 131–148; Mizuko Ito and Daisuke Okabe, "Technosocial Situations: Emergent Structurings of Mobile Email Use," in *Personal, Portable, and Pedestrian: Mobile Phones in Japanese Life*, eds. Mizuko Ito, Daisuke Okabe, and Misa Matsuda (Cambridge, MA: MIT Press, 2005), 257–273; Mizuko Ito, "Mobiles and the Appropriation of Place," *Receiver: Mobile Environment*, 8 (2003), http://www.receiver.vodafone.com (accessed 15 February 2006).
69. Yi-Fu Tuan, *Space and Place: The Perspective of Experience* (Minneapolis, MN: University of Minnesota Press, 1977), 6.
70. Michel de Certeau, *The Practice of Everyday Life*, trans. Steven Rendall (Berkeley, CA: The University of California Press, 1984), 117; this specific formulation of place has been taken by others, most notably by Margaret Morse, "An Ontology of Everyday Distraction: The Freeway, the Mall, and Television," in *Logics of Television: Essays in Cultural Criticism*, ed. Patricia Mellencamp (Bloomington, IN: Indiana University Press/London: BFI Publishing), 195.

71. De Certeau, 117.
72. De Certeau, 117 (emphasis in original).
73. For extended discussion, see Michael Sheringham, *Everyday Life: Theories and Practices from Surrealism to the Present* (Oxford: Oxford University Press, 2006), 212–247; and, Ben Highmore, *Everyday Life and Cultural theory: An Introduction* (London: Routledge, 2002), 145–173.
74. Highmore, 154.
75. Thrift, "'Us' and 'Them'," 180.
76. Wilken, *Teletechnologies*, 175.
77. Caroline Bassett, "'How Many Movements?' Mobile Telephones and Transformations in Urban Space," *Open*, 9 (2005), http://www.skor.nl/article-2854-en.html.
78. Marc Augé, *Non-places: An Introduction to Supermodernity*, second edition (London: Verso, 2008), 25.
79. Augé, 64.
80. Mark Poster, "Digitally Local: Communications Technologies and Space," in *A Sense of Place: The Global and the Local in Mobile Communication*, ed. Kristóf Nyíri (Vienna: Passagen Verlag, 2005), 33.
81. Thrift, "'Us' and 'Them'," 182.
82. Justin Spinney, "Cycling the City: Non-Place and the Sensory Construction of Meaning in a Mobile Practice," in *Cycling and Society*, eds. Paul Rosen, Peter Cox, and David Horton (Aldershot, Hampshire: Ashgate, 2007), 27.
83. Marsha Berry and Margaret Hamilton, "Changing Urban Spaces: Mobile Phones on Trains," *Mobilities*, 5(1), February (2010), 111–129.
84. Iain Sutherland, "Mobile Media and the Socio-technical Protocols of the Supermarket," *Australian Journal of Communication*, 36(1) (2009), 73–83.
85. Spinney, 26.
86. Spinney, 26.
87. Spinney, 25.
88. Spinney, 29.
89. Kazys Varnelis and Anne Friedberg, "Place: The Networking of Public Space," in *Networked Publics*, ed. Kazys Varnelis (Cambridge, MA: MIT Press, 2008), 15–42.
90. Mahyar Arefi, "Non-place and Placelessness as Narratives of Loss: Rethinking the Notion of Place," *Journal of Urban Design*, 4(2) (1999), 182.
91. Varnelis and Friedberg, 39.
92. Varnelis and Friedberg, 39.
93. Thrift, "'Us' and 'Them'," 182.
94. Thrift, "'Us' and 'Them'," 182.
95. Coyne, *The Tuning of Place*.
96. See, for instance: Nicola Green, "On the Move: Technology, Mobility, and the Mediation of Social Time and Space," *The Information Society* 18 (2002), 281–292; Kristóf Nyíri, ed., *A Sense of Place: The Global and the Local in Mobile Communication* (Wien: Passagen, 2005); Rich Ling and Per E. Pedersen, eds., *Mobile Communications: Re-Negotiation of the Public Sphere* (London: Springer-Verlach, 2005); Joachim Höflich, "Places of Life—Places of Communication: Observations of Mobile Phone Usage in Public Places," in *Mobile Communication in Everyday Life: Ethnographic Views, Observations, and Reflections*, eds. Joachim R. Höflich and Maren Hartmann (Berlin: Frank & Timme, 2006), 19–51; Sharon Kleinman, ed., *Displacing Place: Mobile Communication in the Twenty-First Century* (New York: Peter Lang, 2007); Heather A. Horst and Daniel Miller, eds., *The Cell Phone: An Anthropology of Communication* (New York: Berg, 2006); James E. Katz, ed., *Handbook of Mobile Communication Studies* (Cambridge, MA: 2008);

Larissa Hjorth, *Mobile Media in the Asia-Pacific: Gender and the Art of Being Mobile* (New York: Routledge, 2009); Grant Kien, *Global Technography: Ethnography in the Age of Mobility* (New York: Peter Lang, 2009); Nicola Green and Leslie Haddon, *Mobile Communications: An Introduction to New Media* (London: Berg, 2009).

97. For instance: Eric Gordon and Adriana de Souza e Silva, *Net Locality: Why Location Matters in a Networked World* (Boston: Blackwell-Wiley, 2011); Christian van't Hof, Rinie van Est, and Floortje Daemen, *Check In / Check Out: The Public Space as an Internet of Things* (The Hague: Rathenau Institute; Rotterdam: NAi Publishers, 2011): Adriana de Souza e Silva and Jordan Frith, *Mobile Interfaces in Public Spaces: Locational Privacy, Control, and Urban Sociability* (New York: Routledge, 2012).

98. Joshua Meyrowitz, "Global Nomads in the Digital Veldt," in *Mobile Democracy: Essays on Society, Self and Politics*, ed. Kristóf Nyíri (Vienna: Passagen Verlag, 2003), 95.

99. David Seamon, "*A Geography of Lifeworld* in Retrospect."

100. Ingrid Richardson, "Pocket Technospaces: The Bodily Incorporation of Mobile Media," in *Mobile Phone Cultures*, ed. Gerard Goggin (New York: Routledge, 2008), 75; see also, Ingrid Richardson, "Mobile Technosoma: Some Phenomenological Reflections on Itinerant Media Devices," *Fibreculture Journal*, 6 (2005), http://six.fibreculturejournal.org/fcj-032-mobile-technosoma-some-phenomenological-reflections-on-itinerant-media-devices/.

101. Margaret Morse, "An Ontology of Everyday Distraction," 193–221.

102. Caroline Bassett, "'How Many Movements?'"

103. Caroline Bassett, "'How Many Movements?'"

104. Caroline Bassett, "'How Many Movements?'"

105. Joachim Höflich, "A Certain Sense of Place: Mobile Communication and Local Orientation," in *A Sense of Place: The Global and the Local in Mobile Communication*, ed. Kristóf Nyíri (Vienna: Passagen Verlag, 2005), 168.

106. Heidi Rae Cooley, "It's All About the *Fit*: The Hand, the Mobile Screenic Device, and Tactile Vision," *Journal of Visual Culture*, 3(2) (2004), 143.

107. Heidi Rae Cooley, 145.

108. Heidi Rae Cooley, 147.

109. Ingrid Richardson, "Mobile Technosoma."

110. Ingrid Richardson, "Pocket Technospaces"; Ingrid Richardson, "Mobile Technosoma."

111. Ingrid Richardson and Rowan Wilken, "Parerga of the Third Screen: Mobile Media, Place, and Presence" (Chapter 11, this volume).

112. Richardson and Wilken, "Parerga of the Third Screen."

113. David Seamon, "Interconnections, Relationships, and Environmental Wholes: A Phenomenological Ecology of Natural and Built Worlds," in *To Renew the Face of the Earth: Phenomenology and Ecology*, ed. Daniel Martino (Pittsburgh, PA: Duquesne University Press, 2007), 53–86.

114. David Seamon, "*A Geography of Lifeworld* in Retrospect."

115. David Seamon, *A Geography of Lifeworld: Movement, Rest, and Encounter* (London: Croom Helm, 1979).

116. David Seamon, "*A Geography of Lifeworld* in Retrospect."

117. James E. Katz, *Magic in the Air: Mobile Communication and the Transformation of Social Life* (New Brunswick, NJ: Transaction, 2006), 39.

118. Iain Sutherland, "Connection and Inspiration: Phenomenology, Mobile Communications, Place" (Chapter 9, this volume).

119. Esther Milne, *Letters, Postcards, Email: Technologies of Presence* (New York: Routledge, 2009).

120. Kenneth J. Gergen, "The Challenge of Absent Presence," in *Perpetual Contact: Mobile Communication, Private Talk, Public Performance*, eds. James E. Katz and Mark Aakhus (Cambridge, UK: Cambridge University Press, 2002), 227–241.

121. Larissa Hjorth, "Locating Mobility: Practices of Co-presence and the Persistence of the Postal Metaphor in SMS/ MMS Mobile Phone Customization in Melbourne," *Fibreculture Journal*, 6, http://six.fibreculturejournal.org/fcj-035-locating-mobility-practices-of-co-presence-and-the-persistence-of-the-postal-metaphor-in-sms-mms-mobile-phone-customization-in-melbourne/.

122. Naomi Baron, *Always On: Language in an Online and Mobile World* (Oxford: Oxford University Press, 2008).

123. Christian Licoppe, "'Connected' Presence: The Emergence of a New Repertoire for Managing Social Relationships in a Changing Communication Technoscape," *Environment and Planning D: Society and Space* 22(1) (2004), 135–156.

124. Mizuko Ito, "Mobiles and the Appropriation of Place."

125. Rich Ling, *The Mobile Connection: The Cell Phone's Impact on Society* (San Francisco, CA: Morgan Kaufmann, 2004).

126. Emanuel A. Schegloff, "Notes on a Conversational Practice: Formulating Place," in *Studies in Social Interaction*, ed. David Sudnow (New York: The Free Press, 1972), 75–119.

127. Eric Laurier, "Why People Say Where They Are During Mobile Phone Calls," *Environment and Planning D: Society and Space*, 19 (2001): 494.

128. Laurier, "Why People," 495.

129. Laurier, "Why People," 496.

130. Laurier, "Why People," 500.

131. Laurier, "Why People," 500; see also, Ilkka Arminen, "Social Functions of Location in Mobile Telephony," *Personal and Ubiquitous Computing*, 10 (2006), 322.

132. Barry Wellman, "Physical Place and Cyberplace: The Rise of Networked Individualism," in *Community Informatics: Shaping Computer-Mediated Social Relations*, eds. Leigh Keeble and Brian Loader (London: Routledge, 2001), 19.

133. Wellman, "Physical Place and Cyberplace," 29–30.

134. Paul C. Adams, Steven Hoelscher, and Karen E. Till, "Place in Context: Rethinking Humanist Geographies," in *Textures of Place: Exploring Humanist Geographies*, eds. Paul C. Adams, Steven Hoelscher, and Karen E. Till (Minneapolis, MN: University of Minnesota Press, 2001), xiii–xiv.

135. Ann Light, "Negotiations in Space: The Impacts of Receiving Phone Calls on the Move," in *The Reconstruction of Space and Time: Mobile Communication Practices*, eds. Rich Ling and Scott W. Campbell (New Brunswick, NJ: Transaction, 2009), 191–213.

136. Meyrowitz, "The Rise of Glocality," 27.

137. Joshua Meyrowitz, *No Sense of Place: The Impact of Media on Social Behavior* (Oxford: Oxford University Press, 1985), 115.

138. Meyrowitz, "The Rise of Glocality," 26–27. Or, as he puts it elsewhere, "With a greater proportion of our interactions taking place via electronic media, physical copresence is diminishing as a determinant of the nature of interactions." Meyrowitz, "Global Nomads," 96.

139. Varnelis and Friedberg, 39.

140. Varnelis and Friedberg, 32.

141. Adriana de Souza e Silva, "Interfaces of Hybrid Spaces," in *The Cell Phone Reader: Essays in Social Transformation*, eds. Anandam P. Kavoori and Noah Arceneaux (New York: Peter Lang, 2006), 19.

142. See Eric Gordon and Adriana de Souza e Silva, "The Urban Dynamics of Net Localities: How Mobile and Location-Aware Technologies Are Transforming Places" (Chapter 5, this volume).
143. Paul Dourish, "Re-Space-ing Place: 'Place' and 'Space' Ten Years On," paper presented at *CSCW'06* (Banff, Alberta, Canada: November 4–8, 2006), n.p.
144. Francesco Lapenta, "The Infosphere, the Geosphere, and the Mirror: The Geomedia-Based Normative Renegotiations of Body and Place" (Chapter 13, this volume).

2 The Place of Mobility
Technology, Connectivity, and Individualization

Jeff Malpas

The mobile is that which moves; mobility is the capacity to move. Movement can be a movement in place or a movement between places,[1] but without place there can be no mobility.[2] Communication is itself a form of movement—a communing between places—and so carries an essential mobility within it, as well as an essential relation to place. When we talk, in contemporary terms, of "mobile communication," however, we do not refer to the intrinsic movement that all communication exhibits, but rather to an *enhanced* capacity to communicate across changes of place. That this is indeed an enhanced capacity is a reflection of the fact that what has changed with the advent of modern mobility in communication is not a change in the mere possibility for communication across changes of place (there have always been some ways, if often rudimentary and limited, to maintain communication even while on the move), but in the ways in which this possibility can be realized. So great is this change, however, that one might nevertheless say that the change is a *qualitative*, rather than merely *quantitative*, one, and that it constitutes a watershed in the history of communication, marking off the mobile communications of the present, and of the future, from anything that has gone before.

The nature of the break at issue here is very often expressed in terms of a radical shift in the human experience of, and engagement with, *place*, as well as with space and time. Some writers have gone so far as to claim that what is promised by the new technologies, if they have not already achieved it, is a mode of life that is no longer tied to place as once it was. Thus, Barry Wellman suggests that mobile phone technologies enable "a fundamental liberation from place;" he goes on:

> Their use shifts community ties from linking people-in-places to linking people wherever they are. Because the connection is to the person and not to the place, it shifts the dynamics of connectivity from places—typically households or worksites—to individuals. The shift to a personalized, wireless world affords truly *personal communities* that supply support, sociability, information, and a sense of belonging

separately to each individual. It is the individual, and neither the household nor the group, that is the primary unit of connectivity.[3]

The shift from places to individuals that is identified by Wellman might seem to be an obvious consequence of the fact that the mobile phone belongs to a range of mobile electronic devices—personal digital assistants (PDAs), laptops and netbooks, portable music players, handheld game consoles, even portable DVD players (converging in the latest generation of "intelligent" mobiles)—that have proliferated over the last quarter of a century and that are all geared to personal rather than collective use, and that can be carried on or close to the person, being designed for use in no fixed location, and, in some cases, for use while moving. In the mobile world that such devices enable, what remains the same other than the individual that moves? In such a world, in which what is delivered through such mobile devices is itself available and transferable anywhere within a proliferation of networked sites, what is fixed other than the individual who makes use of such devices?

The shift away from place that is associated with the rise of mobile communication technologies is usually taken to be part of a broader shift associated with the character of modernity as such—a shift that occurs in relation to a range of technologies, and not only those of mobile communication, all of which operate through the way in which they change our spatial and temporal relatedness to things, to other persons, and to ourselves, and that is seen to give rise to a world of relationality and interconnection, proximity and compression, changeability and flow. *Globalization* is the term that is most often used to refer to the overall shift that is apparent here. Yet one might argue that although globalization is one aspect of the shift away from place that is characteristic of modernity, the other aspect that must also be attended to, following Wellman, is precisely that of *individualization*. Within the globalizing connectivity of the contemporary world, in which the borders and distinctions between places, times, communities, and even nation-states have become ever so porous and uncertain, the one thing that remains as the basic unit of connection, and through which connectivity operates, is the single *individual*. Such a focus on the individual is neatly captured in what might be thought of as "iBranding"—most obviously Apple's use of product names such as *iPad* and *iPhone*. The "i" may be taken to refer to "information," but it surely also carries connotations of the first person, "I." The *global world* of today is thus also, one might say, *myWorld*—globalization and individualization are brought into one.

Not all have been persuaded, however, of the supposed break with place—whether understood as globalizing, individualizing, or both—that appears to be envisaged here.[4] Joshua Meyrowitz, for instance, has reiterated the fact that human life is always localized, always given over to place. "We are always in place," he writes, "and place is always with us."[5] Meyrowitz does not deny that new communication and information technologies have changed the way we experience place, but insists that these are indeed changes in the experience *of place*, rather than changes that result in the loss of connection to or dependence on place.[6] Meyrowitz's own work can be seen as largely oriented toward understanding the changes in the experience of place that are at work here.[7]

Yet, even if mobile technology, in particular, does not bring about the radical break with place that might appear to be implied by Wellman's analysis, there can be no doubt that contemporary mobile technologies, especially as exemplified in the mobile phone, do indeed tend toward an emphasis on the *individual* rather than the place or the group. In this respect, the mobile phone can often lead us to overlook, or even to forget, the placed (or perhaps "localized") character of our lives. This is not only true of mobile phones, but of modern technology in general. What is evident in the tendency to extol the effects of modern technology in doing away with place, along with the individualizing tendency that accompanies it, can thus be seen to be a feature of the technology as such—its tendency to understand itself as indeed freeing us from the tie to place and as enabling new forms of genuinely "personalized" or globalized connectivity. In this respect, what appears in Wellman's characterization of contemporary mobile technology is part of the very *self-projection* of the technology itself. What makes that self-projection significant, however, is precisely the way in which it remains in tension with what I would argue, like Meyrowitz, is the continuing and inevitable *placedness* of contemporary mobile technology.

Yet what is the role of place here? Why must we suppose that mobile technology, which seems to be able to operate, for the most part, independently of any particular location, nevertheless remains tied to place? The answer, quite simply, is that nothing ever appears except that it appears *in some place*, and so in relation to what stands around it and with which it interacts. Although the same thing may be said to appear in different places, what makes it *the same thing* is precisely the way in which it relates to other things in similar ways—the way, in other words, in which it is always similarly *placed*. Place does not refer us to some mere point on a map or to a kind of neutral container into which things can be put or from which they can be taken out. Place arises in the dynamic interrelatedness of things, even though it is not reducible to any mere system of relations (although on this latter point, I will have more to say later).[8]

To inquire into any phenomenon is also, therefore, to inquire into the place of that phenomenon—and in a literal and not merely metaphorical sense. This is a fundamental principle that applies as much to discussions of mobile technology as it does in any other context. Mobility itself, which is always instantiated in particular places and forms, also carries an essential reference to place. There can be no question, then, of mobile technology, or any other technology, working in a way that does not also implicate notions of place and placedness. The question is not *whether* place is at issue here, but *how* it is at issue. Moreover, any consideration of contemporary technology, and so any consideration of the technology associated with mobile communication, has to take account *both* of the way that technology actually works in place (and so the ways in which it shapes and is shaped by the place in which it functions), *and* the way in which it also brings with it a certain projection of its own mode of functioning, which must include the way in which it projects its own relation to (or separation from) place—which is why the self-projection of contemporary mobile technology is so significant here.

The inevitable and fundamental character of the connection to place is of particular significance when it comes to understanding our own mode of being. Human lives are not lived primarily "in the mind," or even "in the body," but *in the world*—in the places in which we are brought together with other persons and things, and that provide the context for both thought and action. To be "in place" in this way—to be "in the world"—is certainly to be embodied, and, I would argue, to be "minded," but who and what we are is not something determined by the body or even by the mind alone.[9] Put simply, human life is worked out in and through embodiment *in place*, and it is only through such placed embodiment that human lives can appear as meaningful, or, indeed, as human. The structure of embodied "being in place" is thus the basic *ontological* structure within which the *empirical* circumstances of human lives are worked out.[10]

The "working out" of human lives in place operates in a number of different ways and at a number of different levels. Central to understanding that working out, however, is the idea that what we might think of as the "internal" character of human lives is inseparable from the "external" character of those lives—that is to say, the meaning that attaches to attitudes, actions, experiences, and memories cannot be prized away from the worldly objects and events toward which our thinking and acting is directed, from the physical circumstances that give rise to experience as well as to memory, and that also constrain our actions even as they are affected by them, from our very bodies as the means by which we act, by which we feel, and that are the immediate vehicles for self-expression. Our internal lives are thus always externalized, just as the externality of our surroundings is itself reflected back into the internality that we associate with subjectivity.

One might say, in fact, that the subjective always stands in an essential relation to the objective. Consequently, even our sense of self is typically shaped and expressed in terms, not only of our belonging to certain places and locales, as well as to certain communities and groups, but in the very objects that surround us, in the habitual modes and pathways that characterize our movement, in the ways we organize our living and work spaces, in the ways in which we hold our bodies and direct our attention. Because the placed character of human lives is, in this way, also a form of externalization, and because the external, the objective, is that which is accessible *intersubjectively*, so the placed character of human life is itself that on the basis of which human life is to be understood as essentially a mode of being-with-others. We find ourselves and others within the open space that is the world, as that is given in and through our concrete locatedness, and in being so located, we also come to a sense of the world as that within which we stand, individually and collectively, about which we think and speak, and in relation to which we act and experience.

It is within the encompassing structure of place that self, other, and world come together at the same time as they are held apart—it is within place that they are brought into relation and so into differentiation. Consequently, whether our interest is in understanding the larger structure of human being or in examining some particular aspect of human life—such as the operation of some new technological configuration—the focus of our attention must be on the structure of place as that within which the human is given articulation, and the manner in which that structure is altered or differently instantiated. In this respect, the way place already emerges as an issue in the consideration of the technology associated with mobile communication can be seen to intersect with the way in which place appears as the encompassing frame within which the investigation of the impact of such technology operates. In fact, the question concerning the place of mobile communications technology can be understood as identical to the question of how, and in what ways, the place in which human life is constituted and configured is shaped by such technology—to ask after the place of the mobile phone is thus to ask after the way the mobile phone opens up new or different possibilities for human being, which is to say, opens up the place of human being in a new or different way.

One of the very first things to note about the place of the mobile phone, and so also about its place in relation to the human, is that the immediate site of its operation is most often *the hand* (hence the German reference to the mobile phone as a *Handi*) and *the face* (or more specifically the ear and mouth). Moreover, this is not significantly changed by the way such devices are now sometimes used in conjunction with headpieces that no longer require the phone to be held close to the face. Additional to this is the fact that the mobile phone, whether used immediately in the hand or via a headpiece, is nevertheless usually carried on or close *to the body* (and so is available for use in different bodily positions or locations—e.g., when walking or, if often illegally, when driving). This is an important feature of such devices: not only is it part of what makes them *personal* devices, but it is also part of what can lead to them being treated almost as an extension of the person—an extension of the self. Of course, that such devices are an extension of the self is itself a consequence of the externalized conception of the self and subjectivity that I outlined previously, but what is evident in mobile phone use is also a more overt, if partial, identification of self with the external device through the way in which the phone comes to be used and the significance that is ascribed to it.

It is inevitable that different users will relate to devices like mobile phones in different ways that largely reflect existing modes of self-formation—some will remain at a distance from the device, using it only occasionally or in ways that are more straightforwardly instrumentalist, whereas others, especially those who have grown up with it, will relate to it in ways that are much closer, drawing the device much more into the fabric of their everyday activity. For many contemporary mobile phone users, however, the phone has become so much a part of their lives that it functions as an external repository of memory that goes far beyond any old-fashioned diary or calendar, insofar as it contains not just addresses, phone numbers, and dates, but also images, past conversations, and an increasing range of applications to access, manage, and enable a variety of activities and sources of information. In many cases, it also becomes a symbol of one's individual status as belonging to a connected community (or, in the case of those who carry no phone or who use only a very basic model, as not belonging), as the ability to customize ring tones and other aspects of a phone's appearance or operation provides an overt exemplification of the phone as an externalized, if partial, mode of self-articulation.

The mobile phone as material artifact, and so as both a material articulation of self and the immediate, placed focus for mobile communication, often seems to slip away in discussions of mobile communication just as it can easily disappear in its actual usage (that it does so is again tied to the self-projection of the technology).[11] Yet it is important to be reminded of the placed materiality of the device, because it indicates the extent to which mobile communication itself depends on what is given *in place*, as *materially present*, even though what it enables is an engagement *across places*. The reminder of its materiality also reinforces the character of the mobile phone as itself belonging within a structure of externalized, materialized subjectivity articulated in and through place, and so as reinforcing the significance of place, and its necessity, rather than constituting a liberation from it. To reiterate a point already made: the mobile phone, along with the communications technology of which it is a part, is as much tied to place as any other phenomenon, and what is brings about is not a separation from place, but rather a change in the way place is experienced, or better, in the particular way in which place is configured, and the modes of engagement that are operative within it.

The character of the mobile phone as operating within what may be described as the personal space of the body or that is close to the body suggests that whatever changes are brought about by the mobile phone, they will include changes that relate to the space and place of the individual person. Mobile phone use requires that one carry the phone somewhere on or near to one's body, and one typically gives attention to the phone, and to that toward which one speaks and from which one hears, as one also gives primary attention to what one hears and says, rather to that which is around one (more or less the same set of conditions will also apply to the use of other such personal devices). This means that in using a mobile phone one is more focused on what is immediately proximal to one's body than what is at a distance, even a very short distance, away. When one uses a mobile phone within a small, enclosed, or already personalized space, there need be no spatial discontinuity at work—such spaces already tend to be constructed in ways that allow or facilitate an internally directed focus. Public spaces and places, by contrast, are constituted by what is at a distance from the body, not what is immediately proximate. The use of the mobile phone in a public place typically means that the user is turned into their body and away from the place, which also means away from other persons and things in that place, and this has the potential to establish various forms of behavioral and spatial discontinuity. The public place can thus be seen as partially disrupted by the personal place that appears within it.[12] The opposite can, of course, also occur: the fact that the mobile phone can be carried on the person means that mobile communication is possible in situations in which one might expect to be removed from ordinary conversational engagement thereby allowing the intrusion of the interpersonal into the space of the personal.[13]

What is crucial to attend to here is the way the apparent spatial or topographic discontinuity at issue is expressed in behavioral form: the difference in the spatial and topographic engagement is a difference in the nature and possibilities *for action*. This is not to set the behavioral over against the spatial or topographic, but rather to understand the spatial and topographic as always articulated and expressed *behaviorally*. Indeed, the working out of persons in place, and of place in relation to person, is itself a working out that occurs in and through the behavior of bodies in places—which means that to understand a place one must always look to the way behavior is constrained or enabled within it. This is just what we have been doing in looking to the "place" of the mobile phone in the earlier discussion—the discontinuities that emerge are discontinuities in place that are also discontinuities in behavior (often very obvious ones such as walking into other people or things). That we may be distracted, perhaps dangerously so, when we use a mobile device in a public place, and especially when we move through such places, is a simple and familiar point; what is less simple and familiar is the idea that what is at work here is indeed a form of spatial or topographic disruption, and yet that is just what is at issue—and the disruption is not only a disruption manifest in distracted modes of activity, but also in an ambiguity in the very character of the spaces and places concerned.

One of the key points for which Meyrowitz argued in his analysis of the impact of television and other such media on social life was the corrosive effect such media have on our experience and understanding of the differences between various spaces and places, between various spheres of human life and activity, between different aspects of the self and of the social. The tendency at work here seems to be one that is exacerbated under the impact of contemporary mobile communications technology, together with the larger technological frame within which it operates, so that increasingly it seems that the differentiated spaces and places in which human lives are articulated, and from which they can never be entirely removed, appear to break down in the face of a focus on the individual who acts in ways that are often detached from or in tension with, and so may even appear to efface, those spaces and places. In this respect, Wellman's observation that the mobile phone affords "a fundamental liberation from place," and that, within the frame of modern mobile communications technology, it is "the individual, and neither the household nor the group, that is the primary unit of connectivity" picks out a genuine feature of the technology at issue here: although it does not achieve a real "liberation" from place (something that is impossible given the placed character of existence, and especially of human existence), it does bring about a disruption in the character of place, as well as in our own mode of self-understanding, through its emphasis on the individual alone—on the individual as somehow essentially *displaced*.

If we are to understand the nature of the individualizing and displacing tendency that is at work here, then we need to reconsider a feature of place to which I referred briefly in the previous discussion: although places are not reducible to any mere system of relations, places are nevertheless *relational*, arising in the interrelatedness of things, even while they also support such interrelatedness; moreover, places themselves constitute *networks of relations*. Just as anything that appears does so within a context in which it is related to other things—within a place—so does any and every place stand within a larger web of places. Places are thus internally structured in terms of a complex skein of interrelation, at the same time as they are externally connected within and in relation to other places. In this respect, places can be said to fold both inward and outward—outward to other places, inward to the place itself and to that which is given within the place.

The character of place as defined in terms of this combination of inward and outward means, however, that the relational character of place is not the relationality of a simple and evenly distributed "network." The relational character of place is best understood in terms of the relational structure that is evident in horizonality. The horizon marks out a certain field of appearance, drawing together, and so connecting, what appears within the horizon—unifying as it also limits.[14] The horizon does not simply delimit what is given within, but also holds open the possibility of connection to that which is without. The horizon is thus the marker of an internal unity that is also outwardly integrated. That the structure of horizonality should have this character, mirroring the character of place itself, should be no surprise, because horizonality is itself a fundamentally topographical concept.

The idea of relationality is constantly invoked in discussions of contemporary technology. Globalization is itself understood as a form of complex relationality that transcends the usual boundaries, distances, and separations. Relationality is also at work in the idea of connectivity, and in the idea of the individual as the nodal point within the larger system of connections—as Wellman himself says: "It is the individual . . . that is the primary unit of connectivity." Yet the relationality at issue in such talk—in the ideas of globalization, connectivity or individualization—is *not* the same as that which is at issue in the idea of place or horizonality. It is instead the relationality that belongs within a leveled-out and distributed network linking a multiplicity of nodes whose character is exhausted in the relations in which they participate—a relationality that is *quantitative* in character rather than *qualitative*. On such a conception there are no *places* in the sense I have been using it here, but only *positions,* and as such it operates as a purely *spatialized* relationality of the sort exemplified in systems of coordinate geometry. The explicit rendering of such relationality in dynamic terms, terms that often involve notions of trajectory and flow, and that sometimes make appeal to more formal mathematical models, such as those available within dynamical systems theory, does not change the character of that relationality. It remains a relationality that does not work through the heterogeneous, horizontal structure of place, but rather through the essentially homogenous, leveled-out structure of position and so of extended spatiality.[15]

The tension that is evident between the relationality *of place* and what we may call the relationality *of mere position* is often obscured in contemporary analyses by the very prevalence of the language of relationality itself—as if the contrast between a relational and non-relational account was the key issue, rather than the contrast between the placed and the positional. One result of this is that this tension often becomes an unrecognized element, and sometimes a source of inconsistency or obscurity, in those very analyses. It is also a tension that exists in the self-projection of contemporary mobile technologies as somehow standing apart from place at the same time as those technologies continue to operate, as they must, in and through place. Moreover, the mode of self-projection of contemporary mobile technology also carries with it a particular projection of *ourselves,* the human users of that technology, as individuals located purely in terms of our positions within systems of connectivity. In so doing, it projects us in ways that are inconsistent with, and often obscure, our own essential placedness; it projects us in ways that largely reduce who and what we are to the mediated connections and exchanges in which we participate. It is this inconsistency or tension that underlies the behavioral and topographic discontinuities that often arise in conjunction with mobile technology—it also gives rise to discontinuities and tensions in the modes of human life that such technology projects and enables, but which that same technology also obscures.

It is common to think of the connectivity enabled by contemporary mobile and digital technology in terms of an increase in availability and convenience that overcomes the barriers imposed by space, time, and place. We can thus gain access to information, perform certain operations, and connect with other individuals much more readily than we ever could before. Yet, in making other things available to us, and in increasing the convenience and flexibility of our own dealings, we are ourselves made available to others, and to other systems and processes, and so are brought into a system in which we ourselves can be dealt with more conveniently and flexibly—whether this be in our work, in our dealings with government and business, or even in our relations with friends and family. Here is one of the real ironies of the mobile phone: seen as a device that brings "freedom" and individual control, it also brings, by the same token, increased subjection to the technological, and so also the social, political and even economic system of which it is itself a part. The world of mobile communication, of Wellman's "liberation" from place, is thus the same world as that in which we are increasingly made subject to forms of remote surveillance and electronic data collection; in which our behavior is increasingly shaped and directed by bureaucratic management systems both corporate and governmental; in which we are increasingly imprisoned within a network of often electronically mediated consumption and commodification.

Here is the real meaning of the "individualization" that occurs under the impact of contemporary mobile and related technologies: individualization is a transformation in the way in which human beings appear that constitutes them as units within a positional system, and so as available and manipulable within that system. Inasmuch as we appear as simply the nodal points in such a system, so it is easy to lose any sense of oneself, or of one's own sense of place, apart from that system of constant availability and connection. Although it is easy to miscalculate the amount of time dedicated to mobile phone usage (and to other related technologies),[16] there can be no doubt that an increasing amount of everyday activity is given over to forms of behavior that are oriented toward digitally mediated modes of exchange, and that position us within systems of digital connectivity, and that the technology itself projects an image of the increasing prevalence of such connectivity. Our lives thus appear to take the form, more and more, of a continual stream of texts, tweets, messages, emails, blogs, posts, images, transactions—a constant digital chatter in which we are never "alone"—in which there is no real possibility of "solitude"—and yet, precisely as *individualized*, are more alone than we could ever have imagined.[17]

Particular technologies arise only as part of larger technological systems (technology is never just a matter of some single "device"), while technology is itself inseparable from forms of social, economic, and political ordering.[18] The transformation of place into position that mobile technology projects, along with the increasing tendency toward both globalization and individualization that is associated with this transformation, cannot be viewed as merely a phenomenon particular to mobile technology alone. Instead, what we can observe at work in mobile technology is just one manifestation of a much larger and widespread phenomenon: the technological ordering of the contemporary world is an ordering that is centered on the projection of a mode of the world in which there is only position, no *place*, only individuals, no *persons*, and in which *presence* has been reduced to a mere "presentness." It is the world projected as little more than a system of pure connection, relation, and exchange—a technological projection that itself mirrors the system of exchange and connection projected by contemporary forms of economic organization.

Yet as it is a *projection* of a mode of the world, the projection offered by contemporary technology does not erase place or person, nor does it efface any possibility of genuine presence. The positional ordering that contemporary technology projects is itself overlaid on an ordering of place, and it is an ordering on which that positional ordering, and contemporary technology, itself depends. Any "liberation" the mobile phone may offer can occur only through the phone's concrete availability as a material artifact geared to the structure of the human body and to a set of human capacities and desires; the "individualizing" effect of mobile technology operates only in relation to embodied human persons who nevertheless always transcend their character as connected "individuals." It is on the basis of the tension that is evident here between the projection of an essentially displaced mode of ordering and its continuing placedness that the possibility of a genuine critique of contemporary mobile technology opens up (and is already opened up, I would argue, in the work of Meyrowitz and others)—a critique that does not capitulate to the self-projection of the technology itself, and yet also does not lapse into any simple form of anti-technologism.

NOTES

1. The *Oxford English Dictionary* defines mobility as: "Ability to move or be moved; capacity of change of place; movableness . . . Also facility of movement."
2. To respond with the claim that mobility requires only space, and not place, is to misunderstand the way in which even movement in space requires differentiation of the space in which the movement occurs and so requires some analogue of place at the very least.
3. Barry Wellman, "Physical Place and CyberPlace: The Rise of Personalized Networking," *International Journal of Urban and Regional Research*, 25 (2001): 227–257.
4. It is not entirely clear how strongly one should interpret Wellman's talk of "liberation from place." One might argue that Wellman is not claiming that that mobile communication frees us from place altogether, but only that it frees us from the dependence on *specific* places. Still, the rhetoric that is employed here, and not only by Wellman, certainly supports the impression that what is at issue is indeed a quite radical shift, rather than anything more qualified.
5. Joshua Meyrowitz, "The Rise of Glocality: New Senses of Place and Identity in the Global Village," in *A Sense of Place: The Global and the Local in Mobile Communication*, ed. Kristóf Nyíri (Vienna: Passagen Verlag, 2005), 21.
6. See Meyrowitz, "The Rise of Glocality." The persistence of place at issue here is increasingly recognized even by many of those thinkers who may

otherwise be seen as emphasizing a shift to new forms of spatiality—
see, for instance, Nigel Thrift's discussion in "A Hyperactive World," in
Geographies of Global Change: Remapping the World, 2nd edition, eds
R. J. Johnston, Peter J. Taylor, and Michael J. Watts (Oxford: Blackwell,
2002), 29–42.

7. A task already underway in Meyrowitz's ground-breaking work on the social
effects of television and related media in *No Sense of Place: The Impact of
Electronic Media on Social Behavior* (New York: Oxford University Press,
1985). In that work, Meyrowitz identifies the way in which electronic media
change our sense of the differences between places—something that can also
be seen to be at issue, as I discuss later, in the "individualizing" effect of
mobile communications technologies.

8. See, Jeff Malpas, *Heidegger's Topology* (Cambridge, MA: MIT Press, 2006),
for an exploration of the centrality of this idea in Heidegger's thinking, and
so also to philosophical inquiry as such; see also, Jeff Malpas, *Heidegger and
the Place of Thinking* (Cambridge, MA: MIT Press, forthcoming, 2011).

9. One can think of mind and body in a more expansive way as already extending out into the world (as I do in works such as *Place and Experience: A Philosophical Topography*, Cambridge, UK: Cambridge University Press, 1999),
but it is a much narrower conception of both that is commonly assumed—for
the most part we tend to think of the mind as what is internal and private,
what is "inside the head," and the body as that structure of flesh and bone
that is contained within the outer surface of the skin.

10. For the arguments that lie behind this claim, and some account of the structure at issue here, see Malpas, *Place and Experience*.

11. The disappearance of the device in its use is a familiar and frequently noted
phenomenon—perhaps the best-known discussion of the phenomenon is in
Martin Heidegger's *Being and Time*, trans. John Macquarie and Edward
Robinson (New York: Harper & Row, 1962), §§15–16. For my own discus-
sion of the issue, as it relates more specifically to modern digital technology,
see Jeff Malpas, "Acting at a Distance and Knowing from Afar: Agency and
Knowledge on the World Wide Web," in *The Robot in the Garden*, ed. Ken
Goldberg (Cambridge, MA: MIT Press, 2000), 108–125.

12. Although, how this occurs, and to what extent, is clearly dependent on social
and cultural circumstances—the relation between personal and public spaces
in an Indian village street, for instance, is very different from that which
occurs in a North American shopping mall.

13. Such an intrusion of the interpersonal into the personal has developed, however, alongside changes in the size and portability of telephones even prior to
the appearance of mobile and wireless devices. In this respect, the intrusion
of the interpersonal into the personal that is possible with mobile technology is an *intensification* of an existing development rather than something
entirely new.

14. For a discussion of the relation between ground and limit as they operate
here, see Jeff Malpas, "Ground, Unity, and Limit," in *Heidegger and the
Thinking of Place* (Cambridge, MA: MIT Press, forthcoming, 2011).

15. On the distinction between place and space, and so also between place and
"position," see Malpas, *Place and Experience*, 19–43, Malpas, *Heidegger's
Topology*, 258–265, and also Jeff Malpas, "Nihilism, Place, and 'Position',"
in *Heidegger and the Thinking of Place*.

16. In spite of the way in which particular devices disappear in their use, technologies, and particularly consumer-technologies, often present themselves,
in ways that directly thematize their own power and importance; moreover,
as users come to see themselves as dependent on those technologies, so they

may well regard them as having a greater role in their lives than is actually warranted.

17. See, Sherry Turkle, *Alone Together: Why We Expect More from Technology and Less from Each Other* (New York: Basic Books, 2011).

18. For this reason, it is fundamentally mistaken to view technology in primarily *instrumentalist* terms—as if technology could be made to serve certain social or political ends (Kentaro Toyama argues, in "Can Technology End Poverty?," *Boston Review*, Nov/Dec (2010), that mobile technology is less likely to counter existing circumstances and tendencies, than to reinforce and intensify them), but nor should it be viewed as merely an outcome of existing forces—as if it were entirely passive. Technology thus has *effects* that are an outcome of its own character, even though technology is also *embedded* within a larger socio-political framework.

3 Topologies of Human-Mobile Assemblages

Richard Ek

INTRODUCTION

For some years now, media and communication scholars have shown an explicit awareness that spatiality has to be taken into consideration in studying the use of mobile communication technologies. At the center of this discussion is an understanding that traditional notions of space and place as equal to physical distance and physical locality respectively no longer apply or at least have to be complemented or nuanced to a significant extent. The ambition of this chapter is to contribute to this dialogue, and do that in two steps. First, I intend to offer a summary overview of strong social constructionist ideas regarding the conceptualization and characteristics of space and place within human geography, sociological science, and technology studies, usually labeled as Actor-Network Theory (ANT). Here place is crystallized as a relational assemblage of performances, a non-scalar becoming of practices and intensities of agency and stratification. In short, places are social-material events, splayed out in constant flux. Second, I will offer an account of how relational place can be conceptualized differently depending on what kind of ontology is used. This second ambition is executed through a close reading of some seminal articles by the ANT scholars John Law and Annemarie Mol.[1] The theoretical ambition of this chapter is thematically delimited and empirically fleshed out by a focus on topologies of human-mobile assemblages, of the relation between mobile technology—primarily the mobile phone—and the corporeal performance of an individual—embedded in a socio-material network of heterogeneous assemblages.

The chapter is divided into two main sections. The first section starts with a recapitulation of work by media and communication theorists on the need to rethink place as a consequence of the increased use of constantly more sophisticated mobile technologies. The work of Ingrid Richardson and Rowan Wilken is especially attended to here.[2] Then the section continues with the promised summary of conceptualizations of relational place, chiseled out in human geography and ANT. The second section then introduces four topologies developed by Law and Mol: region, network, fluid, and fire. These topologies are introduced to argue the need for an increased ontological awareness when approaching the thematic issues in this anthology. I think this need is urgent indeed! Mobile communications have undergone a radical development under a comparatively short time. We have seen the rise of a multitude of cultural practices that revolve around mobile technology. Mobile technology unfolds new ways of organizing and conducting everyday practices in different spheres of life. Mobile technology and artifacts are more and more apparent and visible. The use of mobile phones and other portable media devices are increasingly a part of the construction of identities and collectivities.[3] Therefore, clearly, there is a need for new and more sophisticated models of analyzing and more complex frameworks of understanding when it comes to the interdisciplinary field of media and communication.[4] In the conclusion, departing from a key argument by John Law, I claim that topological understanding of spatial systems is always political.[5] As a consequence of this, there is a need to also consider the scholarly work on space and mobile technologies as in a wider sense political and reflect upon the implications of this. But there are also, I argue, creative challenges facing researchers working on mobile technologies to actually "think outside the box" (to use a well worn cliché), to think outside the box of secure and established ontologies of the social sciences. Easier said than done, but still a crucial scientific project as mobile technologies increasingly permeate peoples' everyday practices.

RELATIONAL PLACE AND MOBILE TECHNOLOGY[6]

As indicated in the Introduction, the thematic entrance into this issue pivots on the spatiality of an individual and his or her mobile phone. A common interpretation of this spatiality is that mobile technology in general and the mobile in particular works as an extension of the user, as an extension of the user's hand.[7] This echoes McLuhan's view on technology as extensions of embodiment and corporeality, in which the mobile almost becomes an extra organ, always carried along with the body.[8] The increasingly taken-for-granted use of mobile phones in its turn transforms the individual's experience of space and place as the phone enfolds distant contexts into the physically close context of the individual body.[9] Mobile phones are crucial artifacts of mobile technology that lead to altered understandings of place and place-making through processes of ontological renegotiations.[10] Eventually, the mobile phone becomes an artifact that influences how people interact and perceive the societal in general.[11]

The key word here is "mobile," or mobility. Place becomes very much constituted in and through mobility. The person with the mobile phone is a part of the network society outlined by Manuel Castells, subordinated to the logic of the space of flows.[12] The mobility and fluidity of the networked society challenges peoples' experience of place, traditionally very much based on a sedimentary ontology, what Castells labels the logic of the space of place and Heidegger sees as (technologically enframed) dwelling.[13] New forms of daily experience and social interaction are the result, as being-in-the-world increasingly equals being wrapped up within systems of mobile media and communication technologies.[14]

Place here becomes a hybrid between wireless and physical co-presence[15] as the individual and his or her mobile co-produce "movement-place."[16] To Ingrid Richardson, this social trajectory, the individual's increased interaction with a physically distant world, calls for a reimagination and rearticulation of corporeality and materiality. It requires a need to "think through other ontologies, other ways of being-in-the-world, and in a Heideggerian sense, of being-with-equipment."[17] This rearticulation is executed as an analytical schedule: a peripatetic (mobile) post-phenomenology + haptic vision + a view on space and place as relational.[18]

My intention is to discuss the third ingredient further—a view on space and place as relational—but first some few words on the other two analytic variables are needed. The peripatetic post-phenomenology indicates that body-technology relations are contextual and that this contextual body-technology relation is our fundamental ontological condition. Embodiment, in no way determined by the boundaries of the material body, is a capacity that shifts due to the mobile media technologies at hand.[19] Our corporeality extends and withdraws and changes in shape and reach as portable mobile media with prosthetic and orthotic capacities impact upon the body-technology/body-tool relation. Especially the mobile empowers the corporeal schema as the mobile device, due to technological refinement, is an increasingly more sophisticated aural, visual, and haptic interface.[20] Particularly, increased haptic perception (the combination of tactile, kinesthetic, and proprioceptive experiences) becomes a crucial part of the wider phenomenology of technologically mediated mobility.[21]

To Richardson, it is obvious, therefore, that in the teleculture of the twenty-first century, "it is no longer possible to consider space in terms of dichotomized categories of here/there, near/far, personal/private, inner/outer or presence/absence."[22] Instead, there is a need for a relational and variable ontology that imagines and reframes the body as a "material-semiotic assemblage with mediatropic tendencies" and constantly shifting boundaries.[23] It is here that the third ingredient is actualized, as the discussion about places as relation-based and hybrid is founded on such an ontology. This view of place and space is the outcome of a break from the traditional Western metaphysical thought, seeing space as unbounded three-dimensional distance and place as bounded distance. Space has traditionally been seen as a vessel or container in which social and material processes develop. The spatial dimension here has an ontological primacy in relation to the society within.

This primacy has been (and sometimes still is) taken-for-granted and made invisible in the social sciences, but at the same time it has been opposed under the label "the spatial turn." In relational thinking, the ontological primacy is shifted to the objects, processes, events, and so forth, taking place *in* space (from the traditional point of view) and these objects, processes and events (among other phenomenon) *become* space. The notion of space as primarily equal to physical distance and the notion of place as synonymous to a bounded and delimited geographical area still have some relevance in a relational world-view. The crucial difference is that in the relational view on space and place the crucial ontological distinction between objects and space (evident in the framework where objects are placed *in* space) is suspended. The distinction is instead replaced with dialectical frameworks, for instance in the understanding of space as produced.[24]

As a consequence of this ontological shift, places are worked out through social action that changes all the time. Relational place is neither a pure geographical concept (as in the natural sciences) nor a pure social concept (as in some aspatial models in economics) but a specific set of spatialities (including the narratives about them).[25] In a sense, it comes down to the metaphysical choice of being and becoming, the two opposing and enduring presuppositions between Heraclitus (emphasizing the primacy of a fluxing, emergent world) and Parmenides (insisting reality is of a permanent and unchangeable nature).[26] Relational place can only be a place of socio-spatial becoming. This is because relational place is a product of relations, interactive practices and performances. Place and space could therefore be seen as verbs rather than as nouns, each a form of doing that "does not pre-exist its doing [. . .] space is practiced, a matrix of play, dynamic and iterative, its forms and shapes produced through the citational performance of self-other relations."[27] In his call for post-structuralist geographies, Marcus Doel argues that "place is an event [. . .] neither situated nor contained within a particular location, but [. . .] instead splayed out and unfolded across a myriad of vectors [. . .] vectors of disjointure and dislocation [that] may conjugate and reverberate, but there is no necessity for them to converge on a particular experiential or physical location."[28] Relational places are events that are folded into existence, or "actualized" to use a Deleuzian terminology.[29] In sum, there is no space, only spacing, no place, only placing.[30]

In this context, it becomes evident that the idea of geographical scale, indeed any kind of scalar thinking, becomes problematic. Some human geographers have therefore begun to see relational place as co-constituted, folded together, situated, mobile, and multiple. For Ash Amin, this results in a non-scalar and non-linear spatiality:

> [A] topological sense of space and place, a sense of geographies constituted through the folds, undulations, and overlaps that natural and social practices normally assume, without any a priori assumption of geographies of relations nested in territorial or geometric sense.[31]

Every locality becomes an intensity of juxtapositions and intersections of new as well as old spatio-temporalities embedded in layered histories. Relational place is an open, hybrid meeting place,[32] articulated moments in networks of social relations and understandings, but where a large proportion of those relations, experiences and understandings are constructed on a far larger scale that what we happen to define for that moment as the place itself, whether that be a street, or a region or even a continent.[33]

Geographical scales, as a physical framework from the local, through the regional, national, continental to the global, are here inadequate. Rather, scale is an uncertain effect generated by a network and its modes of interaction. It is not scale that counts, but connectivity.[34]

To summarize the characteristics of relational place, it is constituted by relations between objects, subjects, and so on, and not centered on either the objects, subjects, *per se*, nor the "encompassing" space. This ontological shift, the choice of what to prioritize, here indicates a binary thinking with alternatives that exclude each other—just the kind of logic that Richardson wanted to avoid. However, the reader should keep in mind that objects, subjects, and container space still matter as objects, subjects and absolute space, but its importance is downgraded compared to the relations between these objects and subjects. However, sometimes the spokespersons for the idea of thinking space relationally seem to forget that physicality and the Cartesian notion of space very well can be crucial aspects to consider to be able to understand the relational spatiality of a certain phenomena or societal tendency.[35] It is thus important to remember that even if mobile technology compresses time-space through its connective capacities, and in that sense "overcomes" space, the material and physical aspects of this distance-killing capacity can in no way be neglected or down-played in a similar vein as in the early discourse on cyberspace and virtual realities in the 1990s.

RELATIONAL PLACE AS TOPOLOGICAL ASSEMBLAGES[36]

The development of the imagination and conceptualization of space as relational in human geography is a project that has gathered substantial influence from ANT. Prolific ANT scholars have in practice cooperated with human geographers as coauthors and special theme issue participants in journals like *Environment and Planning D: Society and Space*. As a scientific approach that emphasizes the hybridity of the world, ANT offers substantial food for thought in relational thinking of space. This close affinity between human geographers and ANT theoreticians is thus not very surprising but nevertheless something that deserves to be stated explicitly (as it is a project of joint forces of advocates of strong social constructivism with politico-epistemological implications).

Probably the most well known ANT scholar, Bruno Latour has for a long time explored how artifacts affect us and inform our experience of the world. One famous example is his take on the citizen-weapon.[37] The gun and the citizen are here actants that enroll each other and become an actor, performing a program of action, a series of intentions, goals, and steps. The actor "citizen-weapon" kills people, not the gun nor the person per se. The program of action is one of four technical mediations (mediation here being close to Heidegger's notion of enframing).[38] Action is thus not a human endeavor, but a property of an assemblage or association of actants (the second mediation), a property that folds time and space (the third mediation) and crosses the line between signs and things (the fourth mediation). Here, non-humans can express normative statements as well as humans; the example used is the speed bump that makes the normative statement "do not drive too fast here."

ANT thus strives for an ontological position between structuralism and agency, stressing the ontological symmetry within a hybrid collective, between humans, non-humans (other living beings), and objects like texts, materials, and so on. It is the assemblages of humans and non-humans of different kinds that really accomplish something like making social structures last and act at distance and in a way that means we can say that "society" is durable and doable thanks to all the non-humans. In other words, society is held together (stabilized, predicable, etc.) not only by social relations between people but (perhaps even more) relations between artifacts (e.g., cyber communities cannot be maintained without the presence of computers, software, communication infrastructure, etc.).[39] Actors like the "citizen-weapon" or the "citizen-mobile" are shaped in these hybrid collectives as they are co-constructed in the networks, not outside the networks.[40] It is the relations that perform agency.[41] And because action of human-mobile assemblages is constructed within the networks, space and place are made out of the relations as well, made out of the materials that are brought together by the assemblage. Mobile places are in practice a network effect, specific space-time configurations conditioned by the relations in and rationalities of the network.[42]

However, in the beginning of the 1990s, ANT was efficiently criticized by Nick Lee and Steve Brown. To them, ANT proclaimed and regarded networks as an all-encompassing ontology and narrative and consequently erased all space for alterity and excluded all Otherness.[43] Sometimes under labels like "ANT 2.0" and "ANT and after," there has been an endeavor to open up to a spatial imaginary more ontologically complex than "ANT 1.0."[44] John Law and Annemarie Mol are two ANT scholars who have especially tried to respond to this critique, investigating the spatialities of several topological constellations beyond the Cartesian logic of Euclidean space. The starting point is the preposition that objects are an effect of networks and, further, that the enactment of objects is simultaneously an enactment of spatial conditions.[45] What counts as a research object is up to the researcher. Usually, an object is something that is physically constituted and thus takes up space or volume in absolute space, but an object may also be alcoholic liver disease[46] or something else that is seemingly ungraspable. We can say that the project initiated by Law and Mol is an effort to widen the ontological register that traditionally has been dominated by the topographical viewpoint based on the Cartesian imagination of space as unbounded three-dimensional distance and place as bounded locality. The topography of Euclidean space (with an emphasis on the demarcation and containment of space as well as its relief) must lose its primacy to unfold an ontologically erased complexity and heterogeneity.

Mol and Law start out by declaring that Cartesian topography is just one form of social topology among others. Topology—emphasizing properties, relationships, interactions, and relations between relationships[47]—is originally a branch of mathematics that here has been adopted and used in a non-mathematical way (the word is also used, albeit a bit differently, in the geo-philosophy of Deleuze and Serres). In mathematics, topology is about the character of objects in space.[48] To Law, it is about the continuity of an object's shape, even if the object is stretched and bended. Topology concerns "the properties of geometric figures which remain invariant under bending, stretching, or deforming transformations, that is, transformations which do not create new points or fuse existing ones."[49] Topology is thus not limited in the same way as topography, because it does not comply with Euclidean restrictions like, primarily, three-dimensionality. In sum, Law and Mol discuss four topologies.[50] The first three involve regions, networks, and fluidity:

> "The social" doesn't exist as a single spatial type. Rather, it performs several kinds of space in which different "operations" take place. First, there are regions in which objects are clustered together and boundaries are drawn around each cluster. Second, there are networks in which distance is a function of the relations between the elements and difference a matter of relational variety. These are the two topologies with which social theory is familiar.[51]

The first social topology, the region, is a "striated, Cartesian geographical space with demarcated zones and co-ordinates. It is the topography of the social in terms of differentiated regions, e.g. structures, systems, or fields based on territorialized and non-transformable objects."[52] The spatiality of this social topology is the Euclidean conception of absolute and relative space. The second social topology, the network, is the social topology that according to Lee and Brown tended to develop into another grand narrative that claimed to speak for everyone and everything (everything and everyone is a part of a network).[53] Here, compared to the social topology of the region, proximity is measured in how the elements of the network hang together, and is not measured in physical distance. This social topology thus harmonizes with a view on space as relational, as discussed earlier. Chris Chesher draws up such a social topology of network-based spatiality in his description of the use of mobiles at a U2 concert. Here, mobile artifacts of technologies transform a specific place (the concert arena) into a hybrid space of sociality and mediation. The mobile phone is used in different coordinated performances but also in improvised ways relating to individual expression.[54]

To break up the dominance of these two social topologies, Mol and Law present the third social topology—a fluid topology:

> Sometimes, we suggest, neither boundaries [emphasized in the topology of the region] nor relations [emphasized in the topology of networks] mark the difference between one place and another. Instead, sometimes boundaries come and go, allow leakage or disappear altogether, while relations transform themselves without fracture. Sometimes, then, social space behaves like a fluid [. . .]. We're looking at variation without boundaries and transformation without discontinuity. We're looking at flows. The space with which we're dealing is fluid.[55]

Fluidity signals the absence of clear definitions in the relations or in the shape of the enrolled elements, because the spatial relations are relentlessly becoming, shifting, and moving. The fluid space of the third topology is a realm of mixtures, generated by robust objects themselves changeable and not well defined, where there are not necessarily clear boundaries. The objects of fluid space are fluid objects of a fluid spatiality that co-exists with the prior two topologies outlined.[56] Mol and Law finish their paper by concluding that the study of fluids will be a study of the "relations, repulsions and attractions which form a flow."[57] In a follow-up article, Law suggests that in fluid space, no particular structure of relations or particular boundary around an object is privileged (actually, mobile boundaries are needed for objects to exist in fluid space). Law also concludes that continuous and smooth change is a precondition for homeomorphism (topological isomorphism, that is a correspondence between two geometric objects). Radical or abrupt change may very well disintegrate or break a fluid object.[58] Gerard Goggin charts such a fluid topology in his book-length investigation of global mobile media. He approaches the mobile as a research object situated and entangled in a wider communication technology context. Processes of technological convergence and divergence, disassembling and reassembling juxtapose and overlap in this wider context and consequently influence the form and function of the mobile. Traditional mobile phones are transformed into new objects as smartphones and iPhones, functionally as well as symbolically.[59]

To make a quick summary of the argument thus far, the key to understanding the multiplicity of space and place lies thus in a multi-ontological approach toward the research object and not only ontology/social topology as the "network," for instance. Objects can thus be ontologically understood as intersections between spatialities that in their turn enact different social topologies as Euclidean, network, and fluid topologies.[60]

But Mol and Law do not stop here. The fourth topology, discussed in a seminal paper in 2001, is outlined to some degree in contrast to the flow topology (with its elemental affinity to water): fire space.[61] In the fire topology, shape is achieved and maintained through the relation between different forms of presence and absence. Object presence depends for instance on simultaneous absence, or rather the simultaneous absence of multiple others. As a consequence it has three attributes:

> [C]ontinuity as an effect of discontinuity, continuity as the presence and the absence of Otherness; and [. . .] continuity as an effect of a star-like pattern in this simultaneous absence and presence: this is what we imagine as the attributes of shape constancy in a topology of fire. Thus fire becomes a spatial formation alongside (and in interference with) Euclidean, network, and fluid space. To say that there is a fire topology is to say that there are stable shapes created in patterns of relations of conjoined alterity.[62]

Together with Vicky Singleton, John Law later clarifies this reasoning. The fire topology is just another topology (among other, not yet discussed topologies), because Otherness is limitless, but is brought forward because it harmonizes with a post-structuralist critique of the metaphysics of presence that is a central point of critique in ANT. Not everything can be brought to presence, and absence is a precondition for presence (and presence is a precondition for absence).[63] *"An object is a pattern of presences and absences."*[64] In this paper, alcoholic liver disease becomes three fire objects or versions, each made differently with their own series of absent presences: in the hospital, in the substance abuse center, and in the general practitioner's surgery. This approach is similar to the hermeneutic interplay between the part and the whole; it becomes the researcher's task to crystallize what is absent but still something that makes a difference regarding the shape and functionality of a specific research object. To Law, for instance, in the accounts of aerodynamic formulas constructed and used to develop airplanes with a capacity to fly faster than sound, the human pilot (in the function of test pilots) as well as the density of the atmosphere are absent presences, invisible in the formalism but still necessary for its actualization.[65] Clearly, which should come as no surprise, this approach from a philosophy of science point of view falls within the camp of strong social constructivism; the world is created through the empirical account rather than that the empirical account is a declaration of an external reality.

To sum up, objects can be imagined in four ways:

> [A]s volumes in Euclidean space; as stable networks of relations, as fluids that gently reshape their configurations; and, finally, as generative links between presences and absences that are both brought, and cannot conceivably be brought together [. . .] implicit here is a commitment to a spatial way of thinking. Objects are shapes that hold their shape, but they do so in four radically different ways: as volumes; as

stable network configurations; as gentle relational reorderings; and as patterns of absent presence.[66]

In an attempt to flesh out this highly abstract, almost cryptic, reasoning I will conclude this section with an empirical study conducted by Julia Pfaff on mobile phones in Swahili trading practices.[67] Focusing on the mobile's mobility she explores how mobiles are exchanged, appropriated and incorporated into the everyday practices of trade and identification. Mobility here comes in two forms, physical mobility as imagined in the social topologies of the region and the network, and mobility in the sense of adaptability and appropriation, the change of the object itself, as imagined in the social topologies of fluidity and fire.[68] Mobility in the social topology of the region, the striated geographical space, is affected as the mobile goes from owner to owner from Canada to Zanzibar and in different locations in Zanzibar as Michenzani and further to Dar es Salaam.[69] The mobile here primarily moves around in Tanzania, a territorialized and non-transformable region with a specific topography (as long as its border is not changed).[70] Mobility in the social topology of the network implies another form of materiality and spatiality, and executes a network space that is an effect of social practices.[71] The mobile here becomes a part of the network space of Tanzanian telecommunications when the SIM card is exchanged from an "international" to a "national" one. But at the same time the mobile has the same shape as before, it has the same function and it looks the same. It has just changed network space, from an international to a Tanzanian telecommunication space.[72]

But it is possible to see the mobile phone as a fluid technological device in liquid space, the third social topology of fluidity. As an object, the mobile phone becomes a technological device in movement in different ways. First, it moves across the region, the first social topology. Secondly, it moves across the network spaces of mobile telecommunication, the second topology. But thirdly, it moves in a functional sense as it simultaneously is transformed, not in a sudden way as it stops working but in a fluid way, as a fluid object. It not only functions as a communication device, but also as an identity creating device, as a cultural symbol of relative wealth, and so on.[73] The portable mobile media device and the mobile human together enact performances that differ functionally as it is fluid in extension and capability, aurally, visually, and haptically. The call for a variable ontology made by Richardson and other media and communication theorists mentioned previously fits well with the ambition by ANT scholars to "understand objects, to characterize and study them, [and . . .] to attend as much to the *mutability* of what lies invisible below the waterline [the realm of fluidity and fire] as to any immutability that rises above the surface [the domains of regions and networks]."[74] When the assemblage of human and mobile is not separated from the practices performed, the person and the mobile together become the fluid object as it changes (ring signal for instance, or changes from being fashionable to becoming unfashionable[75]) in a smooth way that makes it possible for it to continue to function.[76]

Finally, the fire topology indicates, in contrast to the fluidity of the third social topology, abruptness and discontinuous movement and change. Changes are rapid and erratic and a result of forces that are present through absence. Mobility may for instance be possible due to the absence of immobility as in Diken's account of the actualization of revolution due to societal immobility somewhere elsewhere.[77] As a fire object, the human and his or her mobile phone is a pattern of presences and absences. In a network topology this pattern should be the network, but in the fire topology the ontological register is wider (the ontological "radar" also observes what is under the surface) and messy objects and enactments that usually are not fixed in traditional social science methods are revealed.[78] In Pfaff's account, the mobile is not only a technical tool but also a device that co-creates individual expressions and identifications and co-creates boundary flexibility, reachability and transgressions depending on the social circumstances, economic trading activities, and cultural contexts.

These new topological accounts of the mobile and its user thus help us address issues regarding mobile technologies that have been relatively neglected, in the same way as the "circuit of culture" framework in cultural studies makes it easier for Goggin to address cell phone culture and not only a technology or an economic market.[79] Seeing place as relational spatiality reminds us that fixed ontologies, the topographical, territorial world-view, is an ontological trap that is easy to submit to, but must be questioned and compared to other versions of an ontological imaginary. Seeing place and space relationally also directs our interest toward what is not immediately "there," but stills make a difference as it is present through its absence.

What is absent but nevertheless present in the case of Pfaff's account includes, for instance, the colonial history of Africa, the historiography of "Tanzania" and the echoes of colonialism in contemporary globalization, and the creative destruction (or, perhaps more accurately, the destructive destruction) of economies in the name of colonialism and imperialism. Another object (in its own ontological status a fire object) that is present through its absence is AIDS, and yet another is the dependency of processing capacities of raw materials in "Tanzania." Here, the list can go on, because the reasoning involves the work of making associations—which implies that research is normative work that is argued through empirical accounts rather than "proved" empirically—as the strong constructivist school argues.

CONCLUSION

The ambition of this chapter has been to engage with the dialogue conducted by media and communication scholars on spatial ontologies and the need for a spatial understanding that allow notions of space and place to break away from the Cartesian grid. In ontological terms, this implies that place should rather be seen as produced through action and interaction, that it is fluid and cannot be seen as continuous or compartmentalized. The engagement was conducted through a reading of the four social topologies as they are outlined by ANT scholars John Law and Annemarie Mol: regions, networks, fluids, and fire. The two first social topologies are familiar ground in the social sciences. The topology of fluidity has been approach by media and communication scholars like Ingrid Richardson among others, but from another, more post-phenomenological epistemological perspective. The topology of fire, however, as far as I can observe, has hitherto not been touched upon and hopefully this chapter can initiate attempts to do that, even if I can imagine that further clarification will be needed. We have just started to scratch the surface on the theme "fire topologies of human-technology assemblages." Further work could more explicitly address what is increasingly present in the fire ontology but absent if looked upon from some of the first topologies referred to earlier. However, thinking outside the (Cartesian) box is an endeavor that requires creativity. For instance, as the increasingly sophisticated human-mobile assemblages are permeated by the presence of connectivity, we easily feel abandoned as mobile phone users when there are few mails in the inbox, no new short message service (SMS) on the phone, and so on.[80] There are no longer any excuses (such as, "people have tried to reach me, but because I have been on the move I have not been able to be reached"). Even in a world permeated with global and instant connectivity due to sophisticated mobile communication devices, loneliness easily becomes a present feeling (through the absence of calls and messages). The presence of absent people becomes a feeling and a state of mind that should be approached and addressed more systematically than so far has been done.

However, this future project needs some concluding cautionary remarks. As always, using a certain conception of space and place is a political endeavor and spatial imaginations are not removed from ideology.[81] John Law is the first one to agree on this:

> [S]patial systems are *political*. They are political because they make objects and subjects with particular shapes and versions of the homeomorphic. Because they set limits to the conditions of object possibility. Because they generate forbidden spatial alterities. And because—at least in the case of networks—they tend to delete those alterities. Networks, then, embody and enact a politics linked to and dressed up as functionality.[82]

Thus, any future work in this area requires reflexivity and contemplation, not only because ontology is always political but also because any inquiry into these issues also becomes political. As I have argued in an earlier publication on the same theme, the research produced is in itself a part of the production of space and in a text-context interplay something that must be approached also in epistemological terms.[83] This is especially evident in the cases of fluid and fire topologies, where the research object obviously cannot be separated from the researcher and her or his geographical imagination. Imagination is a key word here, as a plurality of ontologies is based on the "imagination of objects in [. . .] different ways."[84] Looking at the world through a plurality of ontologies is as political as looking at the world from a single ontology, the knowledge created out of multiple worlds as political as knowledge created out of a single world.

NOTES

1. Annemarie Mol and John Law, "Regions, Networks and Fluids: Anaemia and Social Topology," *Social Studies of Science* 24 (1994): 641–71; John Law and Annemarie Mol, "Situating Technoscience: An Inquiry into Spatialities," *Environment and Planning D: Society and Space* 19 (2001): 609–21; John Law, "Objects and Spaces," *Theory, Culture and Society* 19 (2002): 91–105; and, John Law and Vicky Singleton, "Object Lessons," *Organization* 12 (2005): 331–53.
2. Rowan Wilken, "From *Stabilitas Loci* to *Mobilitas Loci*: Networked Mobility and the Transformation of Place," *The Fibreculture Journal* 6 (2005); Ingrid Richardson, "Mobile Technosoma: Some Phenomenological Reflections on Itinerant Media Devices," *The Fibreculture Journal*, 6 (2005); Ingrid Richardson, "Pocket Technospaces: The Bodily Incorporation of Mobile Media," *Continuum: Journal of Media and Cultural Studies* 21 (2007): 205–15; Ian MacColl and Ingrid Richardson, "A Cultural Somatics of Mobile Media and Urban Screens: Wiffiti and the IWALL Prototype," *Journal of Urban Technology* 15 (2008): 99–116; Ingrid Richardson and Rowan Wilken, "Haptic Vision, Footwork, Place-Making: A Peripatetic Phenomenology of the Mobile Phone Pedestrian," *Second Nature: International Journal of Creative Media* 1 (2009): 22–41; Ingrid Richardson, "Faces, Interfaces, Screens: Relational Ontologies of Framing, Attention and Distraction," *Transformations* 18 (2010); Ingrid Richardson, "Ludic Mobilities: The Corporealities of Mobile Gaming," *Mobilities* 5 (2010): 431–447.
3. Gerard Goggin, *Cell Phone Culture: Mobile Technology in Everyday Life* (London: Routledge, 2006), 2–4.
4. Gerard Goggin and Larissa Hjorth, "The Question of Mobile Media," in *Mobile Technologies: From Telecommunication to Media*, eds. Gerard Goggin and Larissa Hjorth (New York: Routledge, 2009), 8.
5. Law, "Objects and Spaces," 102.
6. This section is to some degree based on Richard Ek, "Media Studies, Geographical Imaginations and Relational Space," in *Geographies of Communication: The Spatial Turn in Media Studies*, eds. Jesper Falkheimer and André Jansson (Göteborg: Nordicom, 2006), 45–66.
7. Virpi Oksman, and Pirjo Rautiainen, "Extension of the Hand: Children's and Teenagers' Relationship with the Mobile Phone in Finland," in *Mediating the Human Body: Technology, Communication and Fashion*, eds. Leopoldina Fortunati, James E. Katz, and Raimonda Riccini (Mahwah, NJ: Lawrence Erlbaum Associates, 2003), 103–12.

8. Michael Arnold, "On the Phenomenology of Technology: The 'Janus-faces' of Mobile Phones," *Information and Organization* 13 (2003): 246; Marshall McLuhan, *Understanding Media. The Extensions of Man* (London: Routledge, 1964).

9. Adriana de Souza e Silva, "Mobile Networks and Public Spaces: Bringing Multiuser Environments into the Physical Space," *Convergence* 10 (2004): 15.

10. Wilken, "From *Stabilitas Loci*," 10.

11. Berry and Hamilton, "Changing Urban Spaces," 112.

12. Manuel Castells, *The Rise of the Network Society* (Oxford: Blackwell, 1996); McGuigan, "Towards a Sociology," 47.

13. Martin Heidegger, *The Question Concerning Technology, and Other Essays* (New York: Harper & Row, 1977); Arnold, "On the Phenomenology," 232, 236.

14. John Urry, *Mobilities* (Cambridge, UK: Polity, 2007), 31, 45; Anthony Elliott and John Urry, *Mobile Lives* (New York: Routledge, 2010), 5; Martin Heidegger, *Being and Time* (New York: Harper & Row, 1962).

15. Adriana de Souza e Silva, "From Cyber to Hybrid: Mobile Technologies as Interfaces of Hybrid Spaces," *Space and Culture* 9 (2006): 217–25; Rowan Wilken, "Mobilizing Place: Mobile Media, Peripatetics, and the Renegotiation of Urban Places," *Journal of Urban Technology* 15 (2008): 42–43.

16. Nigel Thrift, "Movement-Space: The Changing Domain of Thinking Resulting from the Development of New Kinds of Spatial Awareness," *Economy and Society* 33 (2004): 582–604.

17. Richardson, "Mobile Technosoma," 5.

18. Richardson and Wilken, "Haptic Vision," 30.

19. Richardson, "Mobile Technosoma," 5; MacColl and Richardson, "A Cultural Somatics," 3; Richardson, "Pocket Technospaces," 207; Richardson and Wilken, "Haptic Vision," 23. This particular version of phenomenology is based on the works of Merleau-Ponty and Don Ihde, and, indirectly, Heidegger.

20. Richardson, "Pocket Technospaces," 206, 214.

21. Richardson and Wilken, "Haptic Vision," 29–30.

22. Richardson, "Pocket Technospaces," 212.

23. Richardson, "Mobile Technosoma," 7; Richardson, "Pocket Technospaces," 206.

24. Martin Jones, "Phase Space: Geography, Relational Thinking, and Beyond," *Progress in Human Geography* 33 (2009): 487–506; Henri Lefebvre, *The Production of Space* (Oxford: Blackwell, 1991).

25. Stephen Graham and Simon Marvin, *Splintering Urbanism. Networked Infrastructures, Technological Mobilities and the Urban Condition* (London: Routledge, 2001), 203; Ray Hudson, *Producing Places* (New York: The Guilford Press, 2001), 257.

26. Robert Chia, "Organisation Theory as a Postmodern Science," in *The Oxford Handbook of Organisation Theory. Meta-Theoretical Perspectives,* eds. Haridimos Tsoukas and Christian Knudsen (Oxford: Oxford University Press, 2003), 114–115.

27. Gillian Rose, "Performing Space," in *Human Geography Today*, eds. Doreen Massey, John Allen, and Pierre Sarre (Cambridge, UK: Polity Press, 1999), 293.

28. Marcus Doel, *Poststructuralist Geographies. The Diabolical Art of Spatial Science* (Edinburgh: Edinburgh University Press, 1999), 7.

29. Nick Bingham and Nigel Thrift, "Some New Instructions for Travellers. The Geography of Bruno Latour and Michel Serres," in *Thinking Space,*

eds. Mike Crang and Nigel Thrift (London: Routledge, 2000), 290; Doreen Massey, *For Space* (Cambridge, UK: Polity Press, 2005), 130. See also Jayne Rodgers, "Doreen Massey: Space, Relations, Communications," *Information, Communication and Society* 7 (2004): 273–91.

30. Marcus Doel, "Un-glunking Geography: Spatial Science after Dr Seuss and Gilles Deleuze," in *Thinking Space*, eds. Mike Crang and Nigel Thrift (London: Routledge, 2000), 125.

31. Ash Amin, "Spatialities of Globalisation," *Environment and Planning A* 34 (2002): 389.

32. Doreen Massey, *Power-Geometries and the Politics of Space and Time: Hettner Lecture 1998* (Heidelberg: Department of Geography, Heidelberg University, 1999), 22.

33. Doreen Massey, *Space, Place and Gender* (Cambridge, UK: Polity Press, 1994), 145.

34. Nigel Thrift, "Intensities of Feeling: Towards a Spatial Politics of Affect," *Geografiska Annaler B: Human Geography* 86 (2004): 57–78.

35. Jones, "Phase Space." See also David Harvey, *Spaces of Global Capitalism: Towards A Theory of Uneven Geographical Development* (London: Verso, 2006).

36. This section is primarily philosophical in its ambition to outline a set of topologically different versions of the world of human-mobile assemblages, but with some snapshots of empirical character for illustrative purposes.

37. Bruno Latour, *Pandora's Hope: Essays on the Reality of Science Studies* (Cambridge, MA: Harvard University Press, 1999), 174–215.

38. Latour, *Pandora's Hope*, 178.

39. David Murakami Wood and Stephen Graham, "Permeable Boundaries in the Software-sorted Society: Surveillance and Differentiations of Mobility," in *Mobile Technologies of the City*, eds. Mimi Sheller and John Urry (London: Routledge, 2006), 179. See also, Michel Callon, "Some Elements of a Sociology of Translation: Domestication of the Scallops and the Fishermen of Saint Brieuc Bay," in *Power, Action and Belief: A New Sociology of Knowledge?*, ed. John Law (London: Routledge, 1986), 196–233; Bruno Latour, "The Powers of Association," in *Power, Action and Belief: A New Sociology of Knowledge?*, ed. John Law (London: Routledge, 1986), 264–80; Michel Callon, "Techno-Economic Networks and Irreversibility," in *A Sociology of Monsters: Essays on Power, Technology and Domination*, ed. John Law (London: Routledge, 1991), 132–64; Bruno Latour, "Technology is Society Made Durable," in *A Sociology of Monsters: Essays on Power, Technology and Domination*, ed. John Law (London: Routledge, 1991), 103–131.

40. John Law, "After ANT: Complexity, Naming and Topology," in *Actor Network Theory and After*, eds. John Law and John Hassard (Oxford: Blackwell, 1999), 9; Jonathan Murdoch, *Post-Structuralist Geography: A Guide to Relational Space* (London: Sage, 2006), 67–68.

41. Michael Callon, and Bruno Latour, "Agency and the Hybrid Collectif," *South Atlantic Quarterly* 104 (1995): 481–507.

42. Murdoch, *Post-Structuralist Geography*, 69–70.

43. Nick M. Lee and Steve D. Brown, "Otherness and the Actor-Network: The Undiscovered Continent," *American Behavioral Scientist* 37 (1994): 772–790.

44. Kevin Hetherington and John Law, "Guest Editorial: After Networks," *Environment and Planning D: Society and Space* 18 (2000): 129.

45. Law, "Objects and Spaces," 92.

46. Law and Singleton, "Object Lessons," 331–332.

47. Hetherington and Law, "After Networks," 129; Murdoch, *Post-Structuralist Geography*, 86.
48. Murdoch, *Post-Structuralist Geography*, 100, note 3.
49. Manuel DeLanda, *Intensive Science and Virtual Philosophy* (London: Continuum, 2002), 25–6.
50. Mol and Law, Regions, "Networks and Fluids," 641–71; Law and Mol, "Situating Technoscience," 609–21.
51. Mol and Law, "Regions, Networks and Fluids," 643.
52. Bülent Diken, "Fire as a Metaphor of (Im)Mobility," *Mobilities* 6 (2010): 96.
53. Lee and Brown, "Otherness," 774.
54. Chris Chesher, "Becoming the Milky Way: Mobile Phones and Actor Networks at a U2 Concert," *Continuum: Journal of Media and Cultural Studies* 21 (2007): 217–225.
55. Mol and Law, "Regions, Networks and Fluids," 643, 658.
56. Mol and Law, "Regions, Networks and Fluids," 663.
57. Mol and Law, "Regions, Networks and Fluids," 664.
58. Law, "Objects and Spaces," 99–100.
59. Gerard Goggin, *Global Mobile Media* (London: Routledge, 2011).
60. Law, "Objects and Spaces," 102.
61. Law and Mol, "Situating Technoscience," 615.
62. Law and Mol, "Situating Technoscience," 616.
63. Law and Singleton, "Object Lessons," 342.
64. Law and Singleton, "Object Lessons," 343, my emphasis.
65. Law and Singleton, "Object Lessons," 343. See also John Law, *Aircraft Stories. Decentering the Object in Technoscience* (Durham, NC: Duke University Press, 2002).
66. Law and Singleton, "Object Lessons," 348.
67. Julia Pfaff, "A Mobile Phone: Mobility, Materiality and Everyday Swahili Trading Practices," *Cultural Geographies* 17 (2010): 341–57.
68. Pfaff, "A Mobile Phone," 342, 345. Pfaff, however, does not use the fire topology here.
69. Pfaff, "A Mobile Phone," 346–351.
70. Diken, "Fire as a Metaphor," 96.
71. Diken, "Fire as a Metaphor," 96.
72. Pfaff, "A Mobile Phone," 347.
73. Diken, "Fire as a Metaphor," 96.
74. Law and Singleton, "Object Lessons," 337, original emphasis.
75. Pfaff, "A Mobile Phone," 348.
76. Law and Singleton, "Object Lessons," 338.
77. Diken, "Fire as a Metaphor," 342.
78. Law and Singleton, "Object Lessons," 334. See also John Law, *After Method. Mess in Social Science Research* (London: Routledge, 2004).
79. Goggin, *Cell Phone Culture*, 6.
80. A remark first made by my colleague Ola Thufvesson.
81. Henri Lefebvre, "Reflections on the Politics of Space," in *Radical Geography. Alternative Viewpoints on Contemporary Social Issues*, ed. Richard Peet (London: Methuen, 1977), 341; and, David Harvey, "Between Space and Time: Reflections on the Geographical Imagination," *Annals of the Association of American Geographers* 80 (1990): 432.
82. Law, "Objects and Spaces," 102, original emphasis.
83. Ek, "Media Studies," 56.
84. Law and Singleton, "Object Lessons," 347.

Part II

Media, Publics, and Place-Making

4 When Urban Public Places Become "Hybrid Ecologies"

Proximity-based Game Encounters in *Dragon Quest 9* in France and Japan

Christian Licoppe and Yoriko Inada

The development of ubiquitous computing[1] has stimulated a growing interest in the organization of activities, particularly collaborative ones, accomplished in complex ecologies, where a variety of digital resources, often based on different types of network infrastructures, intersect each other and are simultaneously available. This strand of research, centered on computer-supported collaborative work (CSCW), has focused on the way the "seamlessness" or "seamfulness" of actual situations of use can support collaboration,[2] and how people are able to creatively combine "assemblies" of artifactual resources and rely upon these in order to support new experiences of sociality adjusted to these complex ecologies.[3] The development of locative media seems to go one step further with respect to the traditional version of ubiquitous computing. Their availability is constitutive of "hybrid ecologies" in which different forms of access to a particular place (e.g., through embodied presence and through various screens and terminals) are somehow articulated.[4] This perspective leads to a complete reshaping of our understanding of what was once dubbed "virtual practices" and understood as actions performed online by a de-contextualized user engaging with his networked computer, a kind of practice perhaps best epitomized by online multiplayer games. Recent ethnographic work on the use of games such as that of *World of Warcraft* in China is pushing toward a reconsideration of how such online games are played: "on screen" collaborative actions and co-present interactions are meshed in a way that can only be understood by taking into account the particular "hybrid cultural ecology" of the Internet café or *wang ba*.[5]

Early examples of the uses of locative media concerned various games combining online actions and urban mobilities that led to the construction of "playful urban spaces."[6] Early studies showed how location awareness could support sociality in specific ways, and how the mutual knowledge of proximity made face-to-face encounters relevant.[7] More recently, the idea that various forms of location awareness might support specific forms of cooperation and sociality has led to the notion of "location-based social networks"[8] and an interest concerning their use in urban public spaces to produce "digital cityscapes."[9] It has also made relevant a phenomenological orientation in which researchers try to bring together the articulation of the corporeal experience of mobility, sociality and the location-informed engagement of players with their mobile terminal into a single embodied experience, such articulation work being constitutive of the way players or more generally mobile users may inhabit together "hybrid ecologies."[10]

We can extend this argument to public settings other than the street. As soon as people are equipped with handheld connected digital devices, then any kind of public space, be it the street, the museum or the Internet café, becomes a "hybrid cultural ecology." However, the focus of CSCW research on the way such ecologies support collaboration in general masks a sensitive issue regarding the specific kind of sociality that could be developing in urban spaces *qua* public places. Public places are where strangers can meet one another. The city as a public place is a "world of strangers,"[11] where strangers are ceaselessly experiencing "traffic encounters."[12] Public places are settings in which strangers "appear" all the time[13] and where such "appearances" have therefore to be socially managed in acceptable ways. The normative rules that govern the patterns of such encounters with strangers are an integral part of the "interaction order" that is constitutive of relations in public.[14] As this author has shown, strangers collaborate to move, interact and recognize each other through gaze and embodied resources with an orientation toward minimality and the avoidance of singularizing the other, as in the case of "civil inattention." Moreover, individuals in public places always constitute a potential audience for any events that might occur there.[15] The forms of collaboration that develop in urban public settings are generally marked by a particular form of "laconic urbanity"[16] in which requests for help are, for instance, framed so as to require only limited commitments.

The question is: what happens when urban public places become "hybrid ecologies," these messy and heterogeneous arrays of co-present and digital resources which provide access to simultaneous experiences of location and proximity that must be articulated to accomplish everyday goals? What kind of sociality is supported by these "hybrid ecologies," and what kind of interaction order and public-sphere relations? How do we orient to strangers and how do we "encounter" them in such connected settings? Will the urban hybrid ecologies of the future be as "democratic" as the megalopolis of the twentieth century, in the sense of treating all individuals as equivalent in innumerable traffic encounters? What kind of public place might the digital cityscape, "augmented" with pervasive computing, become?

If we want to understand the use of mobile media in urban hybrid ecologies and the distinctive but culturally shaped way we experience the latter as places, we need detailed ethnographies of their use and of the kind of sociality between strangers and interaction order that evolve around the availability of location-sensitive shared resources. This chapter aims to provide such a case study, through the ethnographic observations we made of the use of *Dragon Quest 9* in France and Japan. *Dragon Quest 9* is a "proximity game"[17] in which the proximity of players becomes a resource in the gameplay. The game scenario provides incentives for players to approach one another, leading them to meet and gather in urban spaces. When this occurs, traditional public places, such as malls, stations or parks, become crowded with groups of players, many of whom are not acquainted with one another (to varying degrees, as we will see), in an ecology which provides them with many different resources to become mutually perceptible and to interact together: those of co-presence (voice, gaze, gestures, etc.), but also those originating from digital networks, available either through a game terminal or a mobile phone. The way players articulate such resources to produce distinctive patterns of relations in public prefigures the kind of interaction order that may develop in public places as hybrid ecologies.

FIELDWORK

We began by studying exchanges posted on several forums used by Japanese players (Mixi and Shitaraba). We were able to discuss online with six players, and interview nine players face-to-face. We also gathered a small corpus of players' electronic discussions for the purpose of documenting some of the phenomena we were investigating. We did the fieldwork in Japan in two waves, in 2009 and 2010. This involved going to the players' gathering points to observe the way they convened in urban public spaces. The places we went to were: Akihabara and Odaiba Fuji TV in Tokyo, and the train station and the Apita commercial mall in Shizuoka, a town about 200 km from Tokyo on the Tokyo-Osaka *Shinkansen* line. In all these places we tried to talk informally with the players and organizers. In France, we observed what happened after the game was commercially issued, and we participated in the first three meetings of players in Paris, in the *Jardin des Plantes* and in the park of Bercy. We also followed electronic conversations between French players on Internet forums, and held face-to-face informal discussions and opportunistic unplanned interviews during the gatherings in which we participated. One of us (Y.I.) also purchased a DS terminal with the French and Japanese versions of the game, enabling him to play in both countries. This was to allow us to familiarize ourselves with this form of play and to acquire thereby a direct experience of proximity-mediated game encounters.

Our fieldwork was therefore as fragmented and multi-sited as the playful activities we were studying: it exploited the various communication infrastructures available to players and tried to move along the various loci of encounter that were relevant to their practice. The seamfulness of playful situations was both a constraint and a resource for our fieldwork, which followed the activity from one site of engagement to another, in line with some recent claims for the development of "multi-sited ethnography."[18]

CONNECTION EVENTS, "APPEARANCES," AND ENCOUNTERS IN *DRAGON QUEST 9*

Dragon Quest 9 as a "proximity game"

Dragon Quest is a game for Nintendo DS terminals. It was initially a *Dragon Quest* game where players perform quests, fight monsters, and gain experience points that enable them to reach higher levels. The ninth version of the game is different in that it has three additional features in the game scenario that exploit the possibility that terminals recognize and connect to one another through a WiFi connection when within 20 or 30 meters, thereby turning the game into a proximity game.[19]

First, players who are within about 20 meters of each other may engage in multi-player gameplay. Second, when within the same range, players may "appear" on each other's screen: a window pops up on the screen with the name of the other player's game character and description (Figure 4.1). When this occurs, the player can go to the place in the game called Rikka's Inn, where the other player's avatar appears. By clicking on it, the player's profile and tag message appear. During the event, some of the resources of the player who has just appeared may get transferred to the initial player. This is the case of maps, for example, some of which are rare and essential to make progress in the game and to reach higher levels. Japanese players call such mediated encounters *surechigai tsushin*. *Tsushin* means "communication" or "transmission" and carries here a connotation of exchange or transaction. *Surechigai* is a substantive form of the verb *surechigau* which combines *sureru,* "to rub" and *chigau,* "to differ." It is ordinarily used to refer to a situation in which the paths of two persons come close (metaphorically in a situation of "rubbing off with one another") before diverging further on. Whether or not the persons notice their proximity during the event is not an issue here. In the case of *Dragon Quest 9* game encounters, it is only the context in which players use this expression that determines whether they describe the event more as a proximity-based screen mediated encounter or as an opportunity for exchange and transaction. This ambiguity reflects nicely the way the game design tends to alloy the experience of mediated proximity with that of the exchange of resources in the game. There is a constraint on the number of *surechigai tsushin* one can experience: the gameplay only allows three connections of this type at any one time. The screen needs to be refreshed before allowing three more to occur, and so on. *Surechigai tsushin* is a crucial feature of the game experience that combines corporeal mobilities and digital actions in a way that may reshape urban experiences. In 2009, a Japanese user might experience ten or more such events during his commute to the center of Tokyo.

Third, there are specific incentives in the gameplay for users to assemble in groups. These go beyond the mere fact that in such gatherings there may be many opportunities for *surechigai tsushin* and exchanging rare game resources. For instance, the game narrative involves an inn (Rikka's Inn in Japan). When several connected players are within a range of 20 meters, more or less of each other, a cellar and even a first floor to the inn may appear where the players can find additional resources to make further progress.

So the game design invites players in many ways to make a strategic use of spatial proximity and its potential serendipitous translation into the event of *surechigai tsushin*. It objectifies such encounters by providing a calculation infrastructure based on the numbering of such events. The number of *surechigai tsushin* a given player has made is made visible in one interface (up to a maximum of 9,999). Japanese players are highly aware of the number of mediated encounters they have made, and the serious players we interviewed took pride in showing us how they had reached this maximum number of 9,999 encounters. This has also led a small firm to make public the global statistics of *surechigai tsushin* per player and per region in Japan, as shown in Table 4.1.

Figure 4.1 The typical range of WiFi-based proximity encounters.

Table 4.1 Statistics of the Spatial Distribution of *Surechigai Tsushin* Events in *Dragon Quest 9*

Department	Number of encounters	Department	Number of encounters
Tokyo	20,837,854	Hyogo	2,422,271
Kanagawa	9,955,226	Kyoto	1,442,264
Chiba	6,131,732	Fukuoka	1,309,693
Osaka	5,453,952	Ibaraki	1,140,374
Aichi	4,006,606	Hiroshima	1,025,484
Hokkaido	2,540,252	Tochigi	1,019,139

Note: This data tracks encounters from July 2009 (date at which this version of the game was made available in Japan) to January 2010. Source: http://member.square-enix.com/jp/special/dq/dqix/census/index3.php

This table shows how the geographic space may be reshaped as a spatial distribution not of individual types of persons or resources, but of particular forms of proximity-based mediated encounters. It shows how ubiquitous computing does not simply deal with the provision of services "on the move," but also enables the constitution of whole calculation infrastructures for connection events, and therefore contributes to a re-temporalization of our representation of space, from static to dynamic geographies of events, leading to an objectification and spatialization of collaborative social events such as encounters, and probably to new ways of apprehending time and space.

MULTIPLE AND HETEROGENEOUS MODES OF "APPEARANCES" IN URBAN PUBLIC SETTINGS RESHAPED AS HYBRID CULTURAL ECOLOGIES

Other *Dragon Quest 9* players may "appear" in several different ways in urban public spaces, because as one plays the game, such settings are experienced as hybrid ecologies, in which different infrastructures for access and communication are made available.

Physical Co-presence

First, other players may appear by just passing one another independently of any screen event, in a way which is indistinguishable from the traffic encounters with equivalent and anonymous strangers that are a commonplace feature of the experience of the modern city. Such encounters unfold as the paths of mobile and anonymous strangers cross. The spatial proximity and visual availability of the strangers who appear thus makes a possible interactional involvement relevant, but within the bounds of an overarching concern of the participants to preserve their "negative face," that is, their "personal territories," their right to remain inattentive and not to be interrupted, and thus their enjoyment of a certain freedom compared to the forms of obligation that the presence of other individuals might produce.[20] So there is constant tension between the way a sudden proximity may make an interaction relevant[21] and the way the urban denizen expects to benefit in such places from "minimal hospitality," which implies a "right to tranquility."[22] One common way to manage this tension is by the means of "civil inattention," which Goffman argues is a characteristic feature of interactions in public places: an exchange of gaze to manage the embodied encounter that acknowledges the mutual proximity but evades any further involvement and singularization of the encounter and its participants.[23] More generally, the particular ecologies of urban spaces, the way people are thrown together there and manage their mobilities and the many occasions in which they rub elbows with anonymous strangers, are constitutive of the way we routinely experience and perform high density urban spaces (city streets, passageways, airports, stations, malls, etc.) as public places.

Ethnomethodologists have used membership categorization analysis[24] to investigate collective emergent patterns of behavior in public places such as queues and the management of files of people crossing each other in the street.[25] Such urban encounters perform city dwellers as mobile and anonymous bodies, endowed with perceptive, expressive and interactional capacities, categorizable as strangers with apparently equal rights and obligations regarding the management of proximities and encounters. Such subjects are usually expected to treat one another on an equal footing and anonymously, in a way which obliterates singularity, that is, all which might connect the person to a singular identity and make that particular identity relevant. The kind of urbanity that is routinely deployed in these public spaces, the enactment of which constitutes a background expectation of proper behavior, performs participants as anonymous and category-less strangers: "by passing from the neighborhood to the city, the definition of urbanity is supposed to rid itself of problems of identity."[26] However, as Goffman has remarked, there are a few particular types of strangers who appear in public with categories attached to them, for example, children, who are supposed to have limited competencies for behaving appropriately in public and for which help-related activities become relevant. Other categories have specialized rights to engage into focused interactions, such as policemen, in which case a policeman/suspect relational pair becomes relevant when the encounter materializes.[27]

Recognizing a fellow *Dragon Quest 9* player in the street is not very easy. There are few outward cues that might show that a stranger nearby is another player, especially in Japan, where a great number of people seem to be engrossed with varying degrees in a terminal of some sort which they are holding in their hand. Therefore when the paths of players who have never met before cross each other and the players come close to one another, if nothing happens on their screens, they will not easily be able to recognize each other visually as fellow players, and they will treat the "encounter" as any other traffic encounter. However, and this is the crux of the matter, things proceed differently when their proximity leads to specific events on their gaming terminal, providing their sudden proximity with many layers of meaningfulness.

"Surechigai-tsushin"

The development of information and communication technologies has allowed for many different ways for known or unknown persons to "appear" or "pop up" in people's ecologies: phone rings, e-mails, instant messaging "notifications,"[28] and so. Such "mediated apparitions" involve the immediate relevance of various categorizations, such as caller/callee or sender/addressee, and, unlike most urban traffic encounters with strangers, the use of identifiers (e.g. phone numbers, Bluetooth identifiers, e-mail addresses, IP numbers, game handles, etc.) that associate the technological infrastructure used with singular persons or terminals within a kind of heterogeneous crowd. Though they turn the persons they are related to into distinct individuals, they are nevertheless often insufficient to allow mutual face-to-face recognition in co-presence.

Ubiquitous computing[29] and particularly communication services involving some degree of sensitivity to proximity introduce new issues into the organization and management of interactions in public places. They hybridize the random screen-based "pop-up" of another player with urban traffic encounters with strangers, precisely because the digital "apparition" only occurs when the paths of the mobile users meet, so that co-present identification and recognition may become possible and, as we will see later, a relevant concern. *Surechigai tsushin* in *Dragon Quest 9* may be considered as a paradigmatic example of this type of hybrid encounter. When the "random encounter" function has been activated on the DS game terminals of two players who get close to one another, the game of the game character of one of them may appear on the game terminal of the former, as in Figure 4.2. This may or may not happen to be reciprocal. For instance, it will not be reciprocal if the player whose avatar "appears" on the terminal of another player has already reached his own limit of three encounters and has not cleared his terminal yet. Whatever may be the case, the player the terminal of which has been "entered" by the other has no way of knowing from the game whether his own character has "appeared" to others, included those who have "appeared" to him.

Moreover, players who have activated their terminal in the random encounter mode may elect to make available one particular map in such mediated encounters, and it will be transferred to the resources of the players encountered without any action being required of the participants to such mediated encounters. The player on the receiving side will just receive a notification that he has been given a map. This allows some players to act as distributors of rare maps. They announce their presence at a given place and time on the forums, and that they will make available a rare map to players who will be near enough for *surechigai tsushin* encounters to occur.

So it is part of the game's attraction that other players may pop up randomly as players get close to one another, and that one's urban mobility may be directly rewarded by screen-mediated encounters and gifts of game resources, based on serendipitous proximities. We did a trial run in Shibuya with our own terminal and experienced rapidly three such "apparitions" of players. Connecting to their avatars in Rikka's Inn we got their tag messages, which typically ran like: "Hello, I am called X (the game name)! I am a legendary brave man well known in Saitama;" "Hi, I live in Tokyo and I am a kaji-tetsudai. My name is X;" "I . . . I . . . am called X. I am zero year old. I am a slow life of the blue sky who lives in the neighborhood." It is remarkable how in these three examples we picked up

Figure 4.2 The game terminal interface showing three "encountered" players.

randomly, the players interweave descriptions related to the gameplay (the name of their avatar and some of its qualities), with references to the place they live, which points toward a world of physical and embodied presence. The design of such tag messages marks an a priori orientation of players toward the hybrid character of *surechigai tsushin*.

Recently, CSCW and ubiquitous computing research have emphasized the need to look at the uses of these technologies outside the laboratory, in ecologies that are unavoidably heterogeneous, "seamful" and messy.[30] The dual concepts of "seamfulness" and "seamlessness" try to capture this.[31] In our case, the seamfulness of the *Dragon Quest* game experience is related to what Goffman calls "evidential boundaries:"[32] onscreen events are not visible outside the screen and are generally not made available to other players nearby. Moreover, when a game encounter occurs on screen there are often no cues to identify which of the individuals nearby is the player who has just "appeared" on one's screen, though such an "apparition" would make such an identification relevant. So there is a potential evidentary boundary separating what goes on on-screen and in co-presence. In such a potentially seamful situation, players are confronted with a potential choice. The first is to ignore or elude the possibility of recognizing the "encountered" player "in real life." This would accomplish the situation as seamful and as unfolding in channels of activity that are maintained as separate. The other possibility is to try to identify the nearby player who has popped up on one's game terminal among crowds of anonymous passersby and potential fellow players. In this case, the player who looks around engages in an effort to make the situation seamless and to align what goes on in her screen with what is happening in her immediate urban surroundings. This shows that seamfulness and seamlessness are not just abstract qualities of a given situation and the hybrid ecology in which it unfolds. They constitute a collaborative and situated practical accomplishment of participants. Furthermore, the fact that several heterogeneous infrastructures co-exist is a dynamic resource in the production of collective behavior in hybrid ecologies. In the case of *Dragon Quest 9*, the heterogeneity of the situation of play is even greater, because players have other resources to interact during and around *surechigai tsushin* events, particularly in Japan.

Internet-based Social Media Interactions

Japanese *Dragon Quest 9* players interact through various Internet-based social media, and particularly three forums:

> *Shitaraba*: This forum is open to all and the messages posted there can remain anonymous. This is felt to be the reason why this website is often the site of aggressive exchanges between players.

Koryakukan: This forum is also open to all, but a pseudonym is required and the brand of the phone used to connect to it is mentioned. This forum therefore offers more cues for identification and recognition than the previous one. One of its functions, which players find particularly useful, allows anyone to create his/her own thread at will.

Mixi: One can only join this forum through the invitation of another member. This website is considered to be the most regulated of the three, and therefore the least prone to what is known as "flaming."

Because the use of mobile Internet is widespread in Japan, players can interact continuously on their mobile phones, not only before and after encounters, but also during them. It is frequent to see players switch from their game terminal to their mobile phone and back, or to hold a terminal in each hand (Figure 4.3).

The mobile Internet communication infrastructure introduces new seams in the *surechigai tsushin* situation, and the possibility of even more fragmented participation frames.

Figure 4.3 A player in the Akihabara meeting spot with the game terminal in his right hand and his mobile phone in the other.

A CONSTITUTIVE TENSION IN THE MORAL
ORDER OF *SURECHIGAI TSUSHIN* EVENTS

Situations of play are seamful and heterogeneous, particularly in Japan, where many players also communicate on the move through mobile web communication channels. The evidential boundaries associated with the availability of three types of communication infrastructure support the collaborative production of "hybrid" participation frames with varied rights and obligations, which require going beyond Goffman's typologies. Seams and evidential boundaries are oriented to as dynamic resources and the seamlessness or seamfulness of the situations of play are not given a priori, but constitute practical accomplishments. More generally, one could say that the degree of seamfulness or seamlessness achieved in a given situation at a particular time, the kind of participation frame that is collaboratively accomplished and made relevant then and there, and the way urban hybrid ecologies are being experienced as public places are three processes that are interwoven and mutually constitutive.

However, there is one dimension of the problem which the concepts of "seams" or evidential boundaries do not describe well. The availability of proximity-sensitive digital communication makes relevant a particular tension in situations of play in public settings, a tension that is related to the issue of identification. Digital communication infrastructures rely on unique identifiers. Whether such identifiers lean on the "technological" side (e.g., IP addresses, Bluetooth identifiers, etc.) or on the "human" side (e.g., handles, tags, pseudonyms, avatars, etc.), they all share the property of singularizing the participants of a digital communication event while allowing them to preserve their real life anonymity: there are no direct links between such identifiers and the usual cues for recognition "in the real life" that might give their identity away. Such identifiers are attached to their "owner" by a string of mediations that allow him or her to remain at a distance and avoid being recognized in person. All this being said, although one would be unable to recognize the strangers one has never met with whom one interacts digitally, should one meet him or her face-to-face, these are not the equivalent strangers whose paths one crosses every day in the street. They have become singularized, made unique by the very communication event they have engaged in and the way it relies on singular identifiers.

With regard to sociality in public places, when people just talk or chat on their mobile phone with others at a distance, there is a complete separation between the persons one may see and talk to in one's surroundings and the remote conversationalist(s). The issue of being able to identify or recognize the remote participant is generally irrelevant. However, this is no longer the case when the digital communication is triggered by proximity, as in the case of *Dragon Quest 9*. Then the "real life" identification of the person with whom one is exchanging digitally becomes not only possible—because there is a good chance that he or she is close enough—but also an issue.

Membership categorization analysis can help clarify the problem. In an event such as *surechigai tsushin*, players are affected by the mobility of fellow players they are unacquainted with. Furthermore, different sets of categories and category-bound activities, with differing rights and obligations, are enacted *simultaneously*. Once another player "enters" one's game space and occupies part of one's screen with his avatar name and tag message, a first set of relational paired categories becomes relevant:[33] the two people are constituted as *Dragon Quest 9* "players," and more specifically "paired players," bound by the *surechigai tsushin* event. This makes relevant various category-bound activities (i.e., playing the game and, more specifically, giving or receiving game resources such as maps) and some types of mutual obligation (such as acknowledging the "gift" and thanking the other player for it), which require for their achievement identification and mutual recognition on one communication channel or another (i.e., in co-presence or through mobile websites).

There is perhaps also a phenomenological side to this. In such an event one player is clearly affected by the actions of a singularized other without being sure that this has been reciprocal. Being affected by the actions of unknown others without the deed being mutually acknowledged and properly managed socially enacts a kind of possible denial, a possible loss of social substance, because one's existence appears as ignored by the responsible party who has affected it. Trying to obtain some form of acknowledgement then appears as a way of obtaining some repair, in order to remain a social being.

Surechigai tsushin events occur mostly in urban public spaces and are triggered by spatial proximity. Therefore, in addition to the categories "(paired) players," they make relevant a second set of categories, usually related to traffic encounters, that of "paired mobile strangers," and a second set of rights and obligations, such as the obligation to manage tactfully the mutual and embodied proximity, and the "right to tranquility" or to keep to oneself, which, as we have already discussed, is characteristic of public space urbanity in industrialized megapoli. So one should avoid anything that goes beyond the minimal kind of acknowledgement characteristic of "civil inattention," including any kind of engagement that might singularize the passing stranger and make him or her different from all the strangers one passes on an everyday basis.

So there is a potential tension that is inherent to the *surechigai tsushin* event itself. As a proximity-triggered digital encounter between fellow players, it implies an orientation toward acknowledging the exchange and identifying the other participant, something which spatial proximity makes possible in real life. As a specialized kind of traffic encounter with mobile strangers in a large city, its tactful management implies an orientation toward avoiding any kind of singularization of the event and its participants, from mutual ignorance to civil inattention. This makes it particularly interesting to make an ethnographic study of the way players manage this potential tension in public places where they may meet and assemble.

PLACES WHERE PLAYERS "MEET" IN JAPAN AND SOME OF THE TYPE OF "ENCOUNTERS" THAT OCCUR THERE

What we will be aiming to do in the next sections is empirically describe some of the behavioral patterns observed when players assemble in public places, and in particular players who exploit the "seams" of the situation to construct original participation frames. Beyond the particulars of the *Dragon Quest 9* gameplay, we are interested here in understanding the kind of interaction order that is supported by public places when these become hybrid ecologies. In the perspective we develop here, places, the kinds of social encounters that unfold there, and the interaction order through which the management of such encounters is made accountable are mutually constitutive. This definition resonates with the approach of some geographers such as Doreen Massey for whom places can only be understood as "articulated moments in networks of social relations and understandings,"[34] that is, as sites of "throwntogetherness."[35] Indeed, in order to understand what an augmented public place might be, we must understand how strangers are thrown together in its hybrid ecology, and how they manage their proximity socially and turn it (or not) into various forms of encounters by relying on the many seams and multiple communication infrastructures of the situation.

We will now describe several types of urban settings in which players meet in Japan, in which assembled groups of acquainted and unacquainted players equipped with proximity-sensitive game terminals and mobile Internet communication resources mingle with crowds of mobile urban denizens.

"Nora" in Akihabara

The first meeting place is famous among Japanese players. It is located just outside one of the main entrances of a large electronics store at the intersection of two busy streets in the high-tech neighborhood of Akihabara. During the day it is loosely delimited with plots and ribbons (Figure 4.4). However, because the wireless connection between game terminals can extend 20 or 30 meters, players can connect standing outside that zone, either from the curb (Figure 4.6), the exit of the store (Figure 4.3) or sitting on a nearby column (Figure 4.5).

When the *Dragon Quest* game was commercialized, a frenzy started in the store. Swarms of players began clogging its aisles and its owner had to

Figure 4.4 Game meeting places.

Figure 4.5 Connecting to games.

try to get the players back out into the street by marking out a place for them to assemble near the exit. The particular location of this meeting place, near the exit of the store and at the intersection of two busy streets, made it a kind of temporary stopping place along dense urban mobility paths, such that not only players but many passersby also go through it.

Weeks and months later, the players went on assembling there and the place became a regularly visited and well-known meeting place among players throughout the country. Players can go there day and night, seven days a week, without having to announce their passage in any way, and be almost sure to find at least a few other players with whom to experience proximity game encounters. According to the time and the day, the socio-demographics of the place vary. Players expect to find more "salarymen" and "office ladies" during the week and more family men and women during the weekend. It is a place where players typically go whenever they have the occasion. They stay for a while, then leave, without any expectation of engaging in co-present or mobile Internet interactions. All this is strikingly epitomized by the biker in Figure 4.6, who visibly and publicly embodies such an orientation toward: (a) the temporary character of his presence; (b) the opportunistic nature of his stay with respect to large-scale mobilities (he does not even dismount); (c) a very limited openness to engage in any form of co-present interaction. His stance makes it clear that his stay is a temporary halt during a larger mobility (he does not get down from his bike) and that he is comfortable with the peripheral position he has, which prevents any form of gazing or verbal engagement.

We made several observations at this place and interviewed a few of the players who were there. Apart from pairs or trios who were acquainted before and had come there together, we did not observe anybody talking to anybody else, or looking around as if trying to identify a particular player. Neither did players try to look at each other's screens either, unless they had come there together. Such behavioral patterns confirm our description of this place as having a kind of interstitial status, as a place in between mobilities or a "place between places" in which one spends a moment to perform *surechigai tsushin* events with anonymous players and without engaging any further.

In the interviews, many players mentioned that one had to be particularly brave to speak to strangers in the street, including players, particularly in Tokyo. Regular visitors there could only remember one or two occasions in which they had spoken to other players in Akihabara, and in each case the player to which they had spoken belonged to a membership

Figure 4.6 A biker stopping by to play.

category that made such a conversation relevant and justified. They were: (a) children (as Goffman has shown, children are not expected to be fully competent and know all the social ropes, and therefore enjoy particular rights to transgress conventional codes and to be helped in public places); (b) foreigners, for similar reasons to children (we experienced this directly: one of us, who is French and visibly foreign, found it easier than the other, who is Japanese, to approach unknown players and interview them; the presence of a foreigner made questions more relevant and admissible); and (c) players who presented themselves as novices and were asking advice to make some kind of progress in the game. It seems that much more generally, making relevant the standard relation pair "master-novice" is a powerful cultural device to warrant requests to/from strangers in public places in Japan.

Nevertheless, all these instances were described by the players we interviewed as occasional and infrequent exceptions to a more general orientation toward ignoring other players except through game encounters. As one player from outside Tokyo put it, "the place in front of Yodobashi in Akihabara is a place where players who do not know each other gather. The point is to communicate solely through the transmission of data mode, with unknown persons." The fact that one should orient to meeting players as strangers there was further emphasized by the name given by players to this particular place. They called it *Nora* which refers to its being wild, without an owner, unregulated. The *surechigai tsushin* encounters performed there are, moreover, perceived as potentially dangerous. One's gameworld could be contaminated by false data coming from the complete strangers one "meets" through the game in such a place and about whom one has no information at all.

Akihabara therefore gives us a taste of what encounters between complete strangers look like in hybrid urban ecologies: mutual ignorance and/ or civil inattention with respect to co-present resources, mutual ignorance with respect to mobile social media interaction, and a focus on automatically generated, proximity-sensitive exchanges in the gameplay. In such a setting, the *surechigai tsushin* event is managed with an orientation toward exploiting the seamfulness of the situation and ignoring the "standard relational pairs" of membership categories which it makes relevant, such as that of "fellow players" potentially exchanging game resources. In a sense, "*Nora*" in Akihabara exacerbates the orientation toward the minimal, anonymous and equal treatment of strangers that civil inattention exemplifies in a usual urban public setting by translating it within a hybrid ecology. Players in Akihabara do not only behave with strangers in this way, but also with fellow players. This means that they have to actively disregard the fact that they meet there also as players who potentially interact through their terminal in a way which singularizes the participants to the game encounter, and puts into play the possibility of acknowledging this through the resources of co-present interaction.

ASSEMBLING IN OCCASIONAL GATHERINGS:
THE CASE OF THE SHIZUOKA STATION MEETINGS

The Hybrid Ecology of the Shizuoka Gatherings

Because there are many incentives in the gameplay for players to meet, groups of players who live in the same area and interact on Internet-based social media have taken to organize regular meetings in various places. Unlike the Akihabara case, these are not permanent urban fixtures. They are planned and announced on player forums and newsgroups accessible through mobile Internet. They mostly take place in malls, in or nearby train stations, or in food courts in busy commercial areas. This particular choice of location reflects how Japanese players establish a strong relationship between proximity-based game encounters and everyday urban mobilities. Meeting places are chosen at the intersection or in the immediate vicinity of well-trodden urban paths. Such a choice allows players to come by in the course of their own activities, either because it is on their everyday commute in the case of a station, or because they can drop by for a few minutes while doing family shopping in the case of malls.

The meetings at Shizuoka station particularly highlight this connection with mobility. They began with the original practice of a well-known player who called himself the "Legendary Distributor." He would take the train from Tokyo and announce that he would visit various stations along the way at scheduled times, in order to enable other players to wait for him there and perform *surechigai tsushin* with him. Small groups of local players in Shizuoka and its region thus got to exchange on the forums and meet him at Shizuoka station. They became acquainted, and the idea of meeting in these stations took on, even when the "Legendary Distributor" was not coming. The "organizer" usually announces meetings on the three forums mentioned previously, and any player is free to join. Those who regularly participate in such gatherings are often able to recognize one another by sight and to identify each other by their game handles and/or forum aliases, when the two differ. However, they do not usually exchange their real names or mobile phone numbers, maintaining a relatively tight boundary between game-related sites of communication and their other activities as individuals.

In a typical meeting of this type, players are immersed in a public place where many non-playing strangers come and go, and where known and unknown players may make their physical presence acknowledged (e.g., by getting close to the group, sometimes even engaging into a face-to-face conversation, etc.) or not (e.g., by doing *surechigai tsushin* while avoiding getting noticed). They may also communicate by relying on public or semi-public mobile Internet resources. Such a game event is therefore more complex with respect to identification issues and the variety of participation frames it may support than those that occur in Akihabara. We observed and interviewed participants at one such meeting at Shizuoka station (Figure 4.7) and we also had online discussions with a few of them.

Figure 4.7 The start of the meeting, with the core group of regulars.

Figure 4.8 Later on with new players, joining in to exchange maps.

"BRAVE" ENGAGEMENTS AND "TIMID" ENCOUNTERS AT SHIZUOKA STATION

In this section, we will discuss various forms of encounters with unknown players, and the way players present on the site may exploit the layered resources and the seamfulness of the situation in different ways to produce distinctive participation frames. At one end of the continuum is the case depicted in Figure 4.8. An unknown player, accompanied by a friend, has been informed of a meeting on one of the forums on which it has been announced. He walks into the group of players gathered there (it is recognizable as a group of *Dragon Quest* players because the meeting has been announced for this time and place, and all those standing there have a DS terminal in hand) and addresses them verbally in order to make *surechigai tsushin* and exchange game resources. In this configuration, two channels of interaction at least (e.g. co-presence and the game itself) have been opened and somewhat aligned, because the physical appearance of the player who has manifested his presence can now be linked to his game handle. On the basis of his physical appearance, he has become recognizable, for future co-present encounters, as the owner of a particular character, even if his "real" name remains unknown. When the co-presence of players is managed thus, the incoming person makes his "visit" a shared encounter between players and the event is accomplished as a seamless whole.

At the opposite end of the spectrum, we have the case of a lady in her late thirties who had gone to do some shopping with her husband and young son around Shizuoka station. She was already aware through the forum that a meeting was to take place at the station and took the opportunity of her presence there to connect her game terminal while eating with her family at a nearby fast food restaurant, thereby gaining access, through *surechigai tsushin*, to the game space of several of the players gathered there. As her unknown character popped up on the screen of some of the players assembled, none of them noticed or discussed it explicitly. When her meal was over, she left her family and got closer to the group, sitting for a few minutes next to the small Shinto monument visible at the left of Figure 4.8, at which point she became "visible" to the ethnographers, who, unlike the players, were actively scanning the scene for a suitable candidate to account for the "apparition" of the unknown character. Then her husband and son joined her and they disappeared into the station together without providing any visible cue that she had participated in the meeting other than the fact that she was holding a game terminal in her hand. She thus displayed another mode of engagement in the meeting, with two separate participation statuses. One the one hand, she behaved as a ratified and active participant in a game encounter, and on the other hand, she displayed a very loose participation status with respect to the co-present encounter. This falls somewhat in between Goffman's categories with respect to the management of co-presence: while she was more engaged than a simple passerby, as demonstrated her particular form of displacement, which took her around the boundaries of the group of players, she was no more engaged than a bystander, because she did not provide any further visible form of orientation toward the gathering as such. Players have a term to describe players displaying such behavior, which, incidentally, is considered perfectly admissible and legitimate. They call them "timid" players. "Timid" players exploit the "evidential boundaries" of the situation in order to maintain a separation between game and co-present encounters, together with a distinct set of rights and obligations. In a "timid" encounter, the situation is accomplished as seamful.

The degree to which this may be done is a collaborative accomplishment, whatever the looseness of the relationship between the "timid" player and the standing group of regulars may be. For the latter, remarking that a new character has popped up on their screen is always a possibility, a kind of "safe topic" in their ongoing and open state of interaction. Though they often do not mention such events, they occasionally do so. Such "noticings" turn the *surechigai tsushin* event with the unknown player into a public matter of discussion within the group. However, they rarely try to look around to spot a person they can link the new character to, unless the character is related to a player they are "acquainted" with (in the sense of interacting regularly with him or her through one medium or another). This is not just passive indifference, for they often justify themselves for not doing so (c.f. below), but an active indifference, akin to civil inattention, which somehow "resists" a normative orientation toward acknowledging digital interactions in co-presence. "Timid" encounters are a collaborative accomplishment which display a particular format of civil inattention, adapted to, and characteristic of urban places augmented as hybrid public places. However, this does not exhaust the possibilities of exploiting the evidential boundaries of the situation to produce even more layered participation frames.

LAYERED "TIMID" ENCOUNTERS AND FRAGMENTED PARTICIPATION FRAMES

An interesting case is the one in which two players who have occasionally exchanged on a forum make arrangements to perform "*surecshigai tsushin*." This "rendezvous" of sorts, is of a special nature, as we will see, for it does not necessarily involve meeting or talking in person. For now, let us discuss in more detail a typical example in which the participants (publicly) discuss their encounter on a forum using their mobile Internet access. From a co-present perspective, they are strangers who do not know one another's real name, have never met in person, and are unable to recognize one another visually. But from a mobile Internet perspective, they are electronic acquaintances: they "talk" together and know each other's pseudonym, which singularizes them. Finally, within the game, they are strangers all over again, for they have never met in it before and are unaware of the names of the characters they use.

We will now look more closely at part of their electronic conversation, which encompasses their strange kind of "rendezvous:"

1. Rick (4–9-2009) 18:40: *Tonight I am going to Aeon Hamamatsu Ichino at 20:00. Since I have many things to do I will communicate with you via forum or messenging. Won't Sugimaro be there?*

2. Baba (4–9-2009) 19:31: *Great! I will also go to Ichino around 20:30 or 21:00. It would be nice if we could do a surechigai. I think I will stay mostly in the food court. I hope it will work well.*

3. Rick (4–9-2009) 20:27: Baba-*san, It's OK ! I am already having a drink in the food court* (^∀^)/

4. Baba (4–9-2009), 21:19: *Rick-san, I've arrived. I'm coming after going around for a while.*

5. Rick (4–9-2009), 23:30: *Baba-san. Hello. We have probably done surechigai easily. You are 55 happy-san aren't you?*

6. Baba (6–9-2009), 12:17: *Rick-san. Thanks a lot for the other day. I think we probably have been able to do the surechigai. I have seen someone who might have looked like Rick-san, but I was not brave enough to go and greet you. I would like to play again another time.*

The arrangements for their encounter start with "co-location work."[36] Rick announces he will go soon (in about one hour) to a place known to be a meeting place for players, and that he will be available only through the mobile Internet channel, because he has other things to do (message 1). Baba responds by announcing she will also go to the mall at about the same time (message 2). She also evokes a possible game encounter in a way which may be heard as a proposal, and then indicates a more precise location in the mall (the food court), thus going one step further in arranging for such a possibility: being together in the mall might not be enough to ensure a game encounter is possible, but being co-present in the food court in the mall probably would be, because of the implied proximity. Rick then announces he is currently in the food court (message 3), to which Baba replies she has arrived (in the mall) but has things to do before joining him (message 4), which projects the game encounter in the future. Messages 3 and 4 show how players use mobile Internet resources on the move to manage practical arrangements and adjust the tempo and timing of their encounter through mobile messaging.

The next message by Rick (message 5) takes place two hours later. It starts with a greeting. We may wonder why the greeting is necessary, given that they were already talking like this way before. This is made retrospectively clear by what immediately follows (i.e., a hypothetical claim that they may have entered each other's screen). So the greeting can be read as oriented toward a change in the mode of presence of Baba, through her likely "apparition" on the screen of Rick's terminal. Rick then ends the message with an identity check that tries to verify whether the character who has entered the game-play is indeed Baba's. This explains retrospectively why he evokes the game encounter with Baba only as a very likely hypothesis: he has no independent way other than Baba's direct confirmation to connect the game character who has appeared with the forum pseudonym "Baba." Such identification work is needed to align what happens on the screen with the online conversation.

Message 5 is left unanswered for more than a day, and the issue is only evoked in the post scriptum of Baba's next message, the main body of which discusses something else (message 6). It first indicates that she has seen (and therefore looked for and noticed) someone who might have been Rick, and then immediately provides an excuse for not engaging in co-present interaction with him (she was not "brave" enough, she says). Such an excuse is oriented to and made relevant by a more general principle, left implicit here, which states that when two persons who are mutually acquainted are very close, their acknowledgement of the fact through gaze or speech is expected.[37] Message 6 therefore shapes retrospectively the encounter as a refined type of "timid encounter," in which the participants avoid physically acknowledging their mutual co-presence, and therefore elect to remain strangers at that level, while acknowledging their proximity through mobile messaging.

THE PARIS CASE (AUTUMN 2010): PARKS AS HYBRID ECOLOGIES

Dragon Quest 9 was commercialized in France in the middle of 2010. In the autumn, a very active player, Bastien, organized a few gatherings of players in Paris, which he announced on web forums and Facebook. They were the first of the kind in France. The game experience in Paris and in Tokyo turned out to be rather different, however, for various reasons. First, the new version of the game had been out only for a few weeks, so French players were only discovering the principle of proximity-based game encounters. Second, there were fewer chances of experiencing unplanned game encounters during commuting in France, because there were fewer players, but also because the use of mobile terminals is not yet as widespread in the Parisian public transport as it is in Japan. In Paris, this is not the kind of everyday occurrence tightly woven into the mobility practices of players it is in Japan. A corollary of this is that French players must assemble on special occasions if they want to have the chance to perform *surechi-gai tsushin*. Finally, the use of mobile Internet resources, though spreading with the current success of PDAs and iPhones, is still much less common in France than in Japan. This means that, while players are usually active on game-related websites, their posts are not usually made on the move from their mobile phone, but from connected computers, that is, before and after the meetings, but not during them, as in Japan. In other words, *Dragon Quest 9* game encounters in France are situations that involve one less communicative layer than in Japan, and, therefore, less opportunities for highly layered participation frames. The choice in France is an alternative: either acknowledging the game encounter in co-presence through gaze or talk, or ignoring it altogether.

MEETING IN PARKS

The first three meetings in Paris were organized outdoors and took place in parks, first in the *Jardin des Plantes*, then in the *Jardins de Reuilly* and then in the park of Bercy. The precise locus for players to meet was loosely indicated by a cardboard figure, with groups of people either sitting (first and second meeting) or standing (third meeting). This was a deliberate choice from the part of the organizer, who wanted to get players out of their rooms, but was wary of losing his independence if he organized meetings near stores, and did not want to be bothered by passersby in places such as stations. This led to a very different configuration than in Japan, where the assemblies of players were set in places highly connected to dense urban traffic.

Parks are places designed for leisure, for immobility, or at least an easy stroll, unlike the hectic motion in streets, stations and malls. Moreover, when players assemble in parks, they tend to aggregate, leaving fewer opportunities for passersby to enter their grouping. The issue of players rubbing elbows with non-players is therefore much less relevant than in Japan, though occasionally one curious walker may enter the gathering or pass near it. The point is that he or she might then appear as moved by curiosity just by doing so.

In such a configuration, some specific work sometimes has to be done to maintain the spatial and social boundaries of the group. Part of this is related to institutional definitions of different kinds of public places. These regulate what may legitimately happen there, leading to possible encounters with institutional representatives (e.g., park wardens and police). In the second meeting, for instance, the French *Dragon Quest* players assembled in the *Jardins de Reuilly* were asked to move. There is a law in France that states that gatherings of more than six persons are forbidden in "enclosed" ("*confiné*") public spaces. The Reuilly gardens being defined as such an "enclosed public space," a group of sixty players or more could not be allowed, and policemen guided the players toward the nearby *Parc de Bercy*, a more "open" public space where such a rule did not apply. Accounts posted afterward on forums emphasized the cooperativeness and the peacefulness of this encounter with the policemen, remarking that the same policemen who had asked the players to move had also volunteered to guide unaware latecomers to the new spot.

However, even in an open public space such as the park of Bercy, boundary work was also required, as shown in our observations of the third meeting. First of all, a large assembly of people standing close together in a public place is a noticeable and accountable event that is legitimately open to the visual and verbal scrutiny of passersby, who may become onlookers. For example, at one point, a group of teenage girls walked by a small stone building on which a few players were sitting (Figure 4.9 and 4.10). One girl broke away from the group and addressed the players:

Figure 4.9   Group of players and the cardboard figure marking the place.

Figure 4.10   Players sitting and standing in the group, and a passerby walking his dog.

1. Girl (to the group): Why do you gather there?
2. One player: It's for a game.
3. Girl (going back to her friends, in a joking tone): Code red alert. We move.

We see that she feels entitled to address the group to ask for an account, and that such a demand is deemed legitimate (an account is provided straightforwardly). The account given by the player refers to the group as gathered to play a (digital) game. It therefore highlights the auto-referential relevance of the category "player." The girl picks up on that category and implicitly on its potential associations, instructing the group of teenage girls to move away in an ironical and disapproving tone which suggests that such associations are unpalatable to the teenage girl that she is. She thus publicly performs herself and her friends as "non-players," and even as the kind of person who prefers to keep away from the group of players for her own good.

There were other occupants of the place with respect to whom boundary work also had to be accomplished. The stone construction on which a few of the players were sitting was also being used by a man. He was laying dishes out to dry there that he had just washed at a nearby tap. He was therefore using the public space of the park for a kind of domestic activity not usually done in a park. It was no picnic either: his bags that were a few meters away, his old clothes and unkempt beard, all suggested that he had no other place to go and might be homeless. After a while, he finally became angry with the players, addressing them aggressively and reproachfully:

Figure 4.11   Stone construction on which several players sat.

Figure 4.12   Group of teenage girls passing by, and a homeless man washing his dishes.

1. Man washing dishes: You have never learned to behave properly? You cannot go somewhere else?
2. A player: It's a gathering of young people.
3. Man washing dishes (while going away): Ah ok *I understand* it's ok I have understood.

The complaints in line 1 are first oriented toward proper behavior and then toward rights of occupancy. It is interesting to note that the player who replies on this occasion makes reference to a different category than the player who had answered the teenage girls. He does not describe himself and the others in the group as "players," but rather as "young people" ("*jeunes*"). It is an interesting choice of category, which shows recipient design and does some work of its own. Had he described the group as "players," this might have shown tactlessness with respect to a recipient who visibly did not have access to consumption-based leisure, and who might even take offence if this were referred to. On the other hand, "*jeunes*" is a category that refers to age, but that could also be heard in this setting as marking a distance between the players and institutional authority as a whole, both because "*les jeunes*" do not yet occupy institutional positions, but also because that category is often associated with an orientation toward transgression. So self-categorizing as a "*jeune*" is a way for the player to display harmlessness with respect to an underprivileged person competing for the occupation of a public space and claiming it as a place to live and do domestic chores in, a use it was not institutionally designed for. Such a categorization seems to do its work, for the homeless man ratifies the account and moves away, thus temporarily relinquishing his claim to occupancy rights (he has also finished washing his dishes) and acknowledging, as the teenage girl did before him, but in a different manner, the invisible boundary between the group of players and the other potential occupants of the park.

A MORE SALIENT ORIENTATION TOWARD THE SEAMLESS MANAGEMENT OF GAME ENCOUNTERS IN SUCH CO-PRESENT ASSEMBLIES OF PLAYERS

Concerning the sociality that develops in such gatherings in France, we observed several significant differences with respect to our fieldwork in Japan. First, it regularly happened that a player got close to another he did not know to cast a look at his screen. Even if this had to be done with some display of tact (i.e., approaching slowly, remaining close for some time, etc.), it shows that, while the screen is also part of the "personal territory" of the French player, access to it is less tightly regulated than it is in Japan. Second, when in the midst of a French gathering, it was common to hear players shout out the name of an unknown character. This is a way: (a) to make a public "noticing" of a *surechigai tsushin* event; (b) to invite the "owner" of the character to self-identify in a public manner (it happened frequently that a player self-selected to reply something like *"that's me"*); (c) to make the owner of the character recognizable in co-presence, so that players who might want to meet him/her in the game might do so by moving toward him/her or addressing him/her directly; (d) to publicly display an orientation toward managing the gathering as a seamless whole, even though many players may still behave "timidly" in the Japanese sense, for instance by ignoring such addresses, if they wish. Such an orientation was also made visible by certain loose comments. For instance, a girl who was sitting on the small stone structure visible in Figure 4.11 turned toward her companion, who was also playing, and said with a voice loud enough for the comment to be heard by nearby players, *"no but it's dumb to play with someone we don't know who it is."*

It is true that it might require slightly less "bravery" to address a stranger in Paris than in Tokyo—though probably not much. However, such normative orientations regarding traffic encounters in urban spaces are made less salient than in Tokyo, due to the particular ecology of players' assemblies in public settings in France. Because they are done in parks, and a tighter boundary is maintained with non-players, it is more relevant for participants to treat unknown persons nearby as committed fellow players (or categorical peers) rather than as strangers, in the sense of urban traffic encounters.

CONCLUSION

Proximity-based games such as *Dragon Quest 9* are a particular case of location-sensitive playful practices. When played in urban environments, proximity games produce a particular form of urban space, shaped as a "hybrid ecology" through the complex kinds of encounter and collaboration they support. Game encounters mix spatial proximity and digital connectivity in a way that involves several "seams" and "evidential boundaries," that is, many resources for the management of multiple identities and the construction of layered participation frames. "Seams" are not an a priori quality of the situation of play but a potentiality that still has to be accomplished practically: players dynamically orient toward the seamlessness or the seamfulness of the situation on a moment-to-moment basis. Finally, the existence and use of multiple communication infrastructures adds degrees of complexity to the forms of engagement into play, as shown with the example of mobile Internet communication resources in Japan.

Playing such games in the street raises the question of how such "playful urban spaces" may also constitute public places, in which participants must openly manage encounters with strangers. In this case, encounters involve two types of unknown persons: urban traffic encounters with non-playing strangers, as in any kind of modern city, and game encounters with (unknown) players. In the latter case, the encounter may preserve the anonymity of the participants, but require that they connect through digital identifiers and pseudonyms. Such forms of connection and interaction therefore unavoidably singularize their participants. Such a singularization raises issues of potential identification and recognition, and creates a potential tension with respect to the urban traffic encounters in which every participant is equivalent to any other, and in which identification is usually avoided. We think this tension is a very general phenomenon, more generally characteristic of the spread of ubiquitous computing in urban spaces and of hybrid ecologies as public settings.

In our case study, game encounters or *surechigai tsushin* make the identification "in real life" a relevant concern for the players involved, and such a normative orientation provides meaningfulness to some of the participation frames the players orient to. The behavior of "timid players," who come to experience game encounters while trying to elude visual or verbal recognition, and also the fact that they are characterized as such on the basis of their particular behavior, provides an example of the consequences of such a tension. It also constitutes a kind of complex form of civil inattention that is specifically adapted to hybrid ecologies in which numerous digital encounters randomly occur. Such a form of "hybrid civil inattention" could therefore become a specific feature of mediated encounters in large and heavily "augmented" cities, as prefigured by "*Nora*" in Akihabara.

Our comparative ethnographic study of the social gatherings of players in Shizuoka and Paris has also yielded intriguing differences in terms of the types of locations chosen. The kind of public place and interaction order assembled players produce in each case, seems highly sensitive not only to national cultures of urban public places, but to two significant factors: (a) the way players and non players may be expected to mingle or to be more or less segregated in a given public place; and (b) expectations regarding the likelihood of prior mutual acquaintance between players gathered there running from the pure strangers/players one expect to encounter in Akihabara to the higher level of familiarity expected in occasional small town gatherings such as the Shizuoka station, with its regulars and its dependence on prior communication on web forums.

If we try to compare our observations in France and Japan, Japanese players closely associate random playful encounters with their everyday mobility. They also elect to gather in areas of heavy traffic (e.g., stations, malls, etc.). The very locations in which they convene are not very distinct from their surroundings, and the boundaries of their assemblies are therefore highly porous to the traffic of urban denizens. This gives the salience to the general norms governing public behaviors in traffic encounters in situations of playful encounters: Japanese players show themselves to be very concerned with preserving the personal territories of other players, by avoiding looking at their screens and addressing them without "good" reason. French players, on the other hand, convene in leisure-oriented public places, such as parks, a step removed from the paths of busy urban denizens, and accomplish a lot of boundary work to minimize the porosity of their gatherings to usual forms of urban traffic. This makes it more relevant to treat close persons as fellow players (in other words, to treat proximity as bound to the relational category of "player") and to relax some of the normative orientations regarding the treatment of strangers in urban public spaces, e.g. to allow much more leeway in behavior, such as looking at another players' screens or talking to them.

Japanese and French assemblies of players in public have therefore evolved different interactional styles in the management of game encounters. These differences are rooted in their respective particular choices of public settings in which to assemble, which also reflect different ways of shaping the game experience: as embedded in everyday activities and mobilities in Japan, and as a leisure activity to be accomplished at a time and place separated from everyday "busyness" in France. These shape the public settings in which players assemble as hybrid ecologies in which the norms of public behavior in public places are not salient in the same way. These hybrid ecologies support the evolution of distinctive interactional styles between players in the management of the game encounters. What is common to both countries, however, is the relevance of a concern for identifying and acknowledging "in real life" fellow players who have appeared on screen (and thus singularized themselves). Such a concern stems from the more fundamental tension we have analyzed, but the way it is oriented to in both countries is shaped by culturally informed choices and values.

We think that all this is much more general, not specific to the *Dragon Quest 9* case. Through this case study, we get a glimpse of the kind of interaction order that appears to characterize an augmented city whose denizens are highly connected in a location-sensitive way. Such an interaction order is not be as "democratic" as the public order in the twentieth-century megalopolis, in which, from Simmel to Goffman, passersby are performed as noticeable but equivalent strangers through civil inattention. It does not revert to the city of earlier times, which was a mix of neighborhoods organized along some form of categorical membership, for instance types of trade or regions of origin. It rather appears as the locus of dense fluxes of mobile denizens in which traffic encounters occur simultaneously at several levels, that of anonymous equivalent mobile bodies, and that of suddenly (through the event of their proximity) and categorically paired individuals through their communication in digital networks, singularized by this connection event. Augmented traffic encounters should therefore be marked by a pervasive concern about the relevance of acknowledging "in the real life" such a digital connection, and a tension between "augmented" forms of civil inattention in which one should ignore such events and the recognition of multiple forms of mutual category-based memberships.

NOTES

1. Mark Weiser, "The Computer for the 21st Century," *Scientific American* 265, no. 3 (1991), 94–104.
2. Matthew Chalmers and Areti Galani, "Seamful Interweaving: Heterogeneity in the Theory and Design of Interactive Systems," in *Proc. ACM DIS 2004* (NY, USA, August 2004), 243–252
3. Jon Hindmarsh, Christian Heath, Dirk vom Lehn and Jason Cleverly, "Creating Assemblies in Public Environments: Social Interaction, Interactive Exhibits and CSCW," *Computer Supported Creative Work* 14 (2005): 1–41.
4. Andy Crabtree and Tom Rodden, "Hybrid Ecologies: Understanding Cooperative Interaction in Emerging Physical-Digital Environments," *Personal and Ubiquitous Computing* 12 (2008): 481–493.
5. Silvia Lindtner et al., "A Hybrid Cultural Ecology: World of Warcraft in China," in *Proc. of CSCW* (San Diego, California, November 2008).
6. Adriana de Souza e Silva and Larissa Hjorth, "Playful Urban Spaces: A Historical Approach to Mobile Games," *Simulation & Gaming* 40 (2009): 602–625.
7. Christian Licoppe and Yoriko Inada, "Emergent Uses of a Location Aware Multiplayer Game: The Interactional Consequences of Mediated Encounters," *Mobilities* 1 (2006): 39–61; Christian Licoppe and Yoriko Inada, "Locative Media and Cultures of Mediated Proximity: The Case of the Mogi Game Location-aware Community," *Environment and Planning D: Society and Space* 28 (2010): 691–709.
8. Adriana de Souza e Silva and Jordan Frith, "Locative Mobile Social Networks: Mapping Communication and Location in Urban Spaces," *Mobilities* 5 (2010): 485–505.
9. Adriana de Souza e Silva and D. M. Sutko, eds., *Digital Cityscapes: Merging Digital and Urban Playspaces* (New York, Peter Lang, 2009).
10. Ingrid Richardson, "Ludic Mobilities: The Corporealities of Mobile Gaming," *Mobilities* 5, no. 4 (2010): 431–447; Ingrid Richardson and Rowan Wilken, "Haptic Vision, Footwork, Place-making: A Peripatetic Phenomenology of the Mobile Phone Pedestrian," *Second Nature* 1, no. 2 (2009): 23–39.
11. Lyn H. Lofland, *A World of Strangers: Order and Action in Urban Public Space* (Prospect Heights: Waveland Press, 1973).
12. Ulf Hannerz, *Exploring the City: Inquiries towards an Urban Anthropology* (New York: Columbia University Press, 1980).
13. Hannah Arendt, *La condition de l'homme* (Paris: Calmann Levy, 1983).
14. Erving Goffman, *Behavior in Public Places* (New York: Free Press, 1963); Erving Goffman, *Relations in Public: Microstructure of the Public Order* (New York: Harper & Row, 1971).
15. Lyn H. Lofland, *The Public Realm. Exploring the City's Quintessential Social Territory* (New York: Aldine de Gruyter, 1998).
16. Isaac Joseph, "L'espace public et le visible," *Architecture & Comportement* 9 (1993): 397–401.
17. Todd Söderlund, "Proximity Gaming: New Forms of Wireless Gaming," in *Digital Cityscapes: Merging Digital and Urban Playspaces*, eds. Adriana de Souza e Silva and Daniel M. Sutko (NY, Peter Lang, 2009), 217–230.
18. George Marcus, "Ethnography in/of the World System: The Emergence of Multi-Sited Ethnography," *Annual Review of Anthropology* 24 (1995): 95–117; Christine Hine, "Multi-sited Ethnography as a Middle Range Methodology for Contemporary STS," *Science, Technology & Human Values* 32 (2007): 652–671.

19. Tom Söderlund, "Proximity Gaming."
20. Penelope Brown and Stephen C. Levinson, *Politeness. Some Universals in Language Usage* (Cambridge, UK: Cambridge University Press, 1987).
21. Christian Licoppe and Yoriko Inada, "Locative Media and Cultures of Proximity."
22. Isaac Joseph, "Activité située et régimes de disponibilité," *Raisons Pratiques* 10 (1999): 157–172.
23. Erving Goffman, *Behavior in Public Places.*
24. Harvey Sacks, *Lectures on Conversation*, 2 vols. (Oxford: Blackwell, 1992); Stephen Hester and Peter Eglin, eds., *Culture in Action. Studies in Membership Categorization Analysis* (Washington DC: International Institute for Ethnomethodology and Conversation Analysis & University Press of America, 1997).
25. Rod Watson, "Some General Reflections on 'Categorization' and 'Sequence' in the Analysis of Conversation," in *Culture in Action*, ed. Stephen Hester and Peter Eglin (Lanham, MD: International Institute for Ethnomethodology and University Press of America, 1997), 49–76.
26. Isaac Joseph, "L'espace public et le visible."
27. Erving Goffman, *Behavior in Public Places*, and *Relations in Public.*
28. Christian Licoppe, "The 'Crisis of the Summons': A Transformation in the Pragmatics of 'Notifications,' from Phone Rings to Instant Messaging," *The Information Society* 26 (2010): 288–302.
29. Mark Weiser, "The Computer for the Twenty-First Century."
30. Genevieve Bell and Paul Dourish, "Yesterday's Tomorrows: Notes on Ubiquitous Computing's Dominant Vision," *Personal & Ubiquitous Computing* 11 (2007): 133–143.
31. Matthew Chalmers and Areti Galani, "Seamful Interweaving."
32. Erving Goffman, *Frame Analysis. An Essay on the Organization of Experience* (New York: Harper & Row, 1974).
33. Harvey Sacks, *Lectures on Conversations*; Rod Watson 1997, "Some General Reflections."
34. Doreen B. Massey, *Space, Place and Gender* (Minneapolis, MN: University of Minnesota Press, 1994).
35. Doreen B. Massey, *For Space* (London, Sage, 2005).
36. Christian Licoppe, "Recognizing Mutual 'Proximity' at a Distance: Weaving Together Mobility, Sociality and Technology," *Journal of Pragmatics* 41, no. 10 (2009): 1924–1937.
37. Christian Licoppe and Yoriko Inada, "Locative Media and Cultures of Proximity."

5 The Urban Dynamics of Net Localities

How Mobile and Location-Aware Technologies Are Transforming Places[1]

Eric Gordon and Adriana de Souza e Silva

INTRODUCTION

Since the popularization of mobile phones, many critics have lamented the decline of public spaces as they have become cluttered with outside connections, drawing positive correlations between global connections and local disconnections.[2] For example, the cultural critic Howard Rheingold observed that passengers on trains and buses in Japan preferred to talk to somebody who was physically absent than with other people who were in the same vehicle.[3] Undoubtedly, mobile phones promote social disruption, especially when used in public spaces such as restaurants and public transportation.[4] For that reason, critical theorist Norman Klein has suggested that when people talk on a mobile phone while walking, they move through physical space, but they are actually not there.[5] Psychologist Kenneth Gergen called this perceived disconnection from physical space "absent presence." For him, absence presence is when "one is physically absorbed by a technologically mediated world of elsewhere."[6]

But despite this criticism, people continue to meaningfully interact with their surroundings. From sidewalks infused with historical significance to plazas that cultivate social interaction, urban spaces serve as platforms for place-making, even as the methods of inhabiting those spaces are expanding to include networked connectivity. It is increasingly common for people to inhabit a hybrid of physical and digital spaces, mediated by the use of mobile and location-aware technologies. This multiple habitation stretches the boundaries of the Aristotelian space typically associated with cities. Digital networks extend the perceived physical context of bodies as a user can be in one space and communicate with someone in an entirely different space. But more importantly, the use of mobile and location-aware technologies in public spaces transforms our experience of places and spaces, creating conditions for the emergence of new spaces. What we call *networked locality* (or net locality) is the cultural and technological framework through which people manufacture places mediated by location-aware and mobile devices and digital networks.

Places have traditionally been defined as networked, dynamic, social, and meaningful entities. Net locality reinforces these conceptions of place, but also alters them in fundamental ways. Shaun Moores[7] criticized Manuell Castells's[8] definition of [the space of] places, arguing that places are not self-contained, because in a city people maintain connections with others who are beyond its physical boundaries. Moores follows Doreen Massey[9] and Joshua Meyrowitz,[10] for whom places are permeable localities composed of networks—that is, they are dynamic entities. Massey argues, "The uniqueness of a place, or a locality, [. . .] is constructed out of particular interactions and mutual articulations of social relations, social processes, experiences and understandings."[11] As Massey suggests, places are also comprised of social relations. Henri Lefebvre[12] does not use the word place, but he addresses how spaces are composed of more than just their physical characteristics: they are socially produced. And David Harvey helps make the connection between Lefebvre's concept of space and the broader thinking about place: "The. . . . Lefebvrian construction permits an examination of the processes of place construction [. . . and it] owes as much to activities in the representational and symbolic realms as to material activities."[13]

In sum, places are the affective framework through which people make meaning of external spaces and incorporate that meaning into their identity, whether or not that space is extended through digital networks. Steve Harrison and Paul Dourish define a place as "a space which is invested with understandings of behavioral appropriateness, cultural expectations, and so forth [. . .] 'Places' are spaces that are valued."[14] The geographer Yi-Fu Tuan suggests, "When space feels thoroughly familiar to us, it has become place."[15] And accordingly, those places get extended out into spaces. As Gaston Bachelard puts it, "the stuff of our 'inner lives' is thus to be found in the exterior spaces or places in which we dwell, while those same spaces and places are themselves incorporated within us."[16]

Net locality is an emerging context with which places are made and communicated. Net localities are also dynamic, meaningful, composed of social interactions and networked connections. In addition, net localities include an increasing amount of location-based information that is not only "attached" to places via location-aware mobile interfaces, but become an intrinsic dynamic part of those places. As Adriana de Souza e Silva and Daniel M. Sutko[17] note, the process of actualizing *information as place* and *place as information*—mediated via location-aware technologies—is an example of how digital information and place, in a constant process of becoming, inform the construction of net locality. Location-aware technologies and the information with which they interface are not outside places, nor is physical material space disconnected from the location-based information now embedded in it. Each is a part of how locations are constructed, occupied, and disseminated.

Net locality has created new contexts for interaction, not all of which are solely determined by being in the same physical location. For example, people can be aware of others' presence through markers on a map or a local tweet.[18] Increasingly, the web is brought into the spaces we occupy, and, similarly, those spaces are brought into the web. But this takes practice. The ability to navigate net localities actually means locating people and information that we associate with our "digital selves" into something of a hybrid space, in which the borders between remote and contiguous contexts no longer can be clearly defined.[19] Net localities are practiced spaces that dynamically develop over time, through social practices with technology. What's more, net localities also include all those people who are co-present in the physical space who are not accessing digital information. The woman walking down the street without a device, not accessing information, becomes part of the situation that comprises the space, because what we experience is actually a qualitative change in our experience of space, rather than just a different type of interaction with the Internet.

Urban spaces have always been mediated by technology. Buildings, cars, street signs—these are all technologies that contribute to the experience of a city street. But net localities produce unique types of networked interactions and, by extension, new contexts for social cohesion. Co-presence is not mutually opposed to networked interaction—and as emerging practices of technology develop, drawing that line in the sand becomes increasingly difficult.

In this chapter, we define net locality as an emerging logic from which places get defined in networked urban spaces. First, following sociologist Georg Simmel, we argue that technologies work as filters that mediate our interactions with urban spaces and to other people in them, by helping us to pay selective attention to some things and not others.[20] Then, we turn to sociologist Erving Goffman to clarify how micro interactions work in and with net localities.[21] This is what we call the net local situation. Finally, we address how the interaction with mobile and location-aware technologies can be understood as a type of performance in public space. Location-aware technologies allow us to seamlessly interact with the remote and the proximate. As such, our goal is to understand net localities so that we might shed light on how to design the technologies that mediate them, as well as to grasp what it is to create new forms of places within networked and information-rich urban environments.

GOOD OL' PUBLIC SPACES

Questions about urban public space have been vigorously debated for at least a century. Simmel noted that the rapid increase in external stimuli found everywhere in the city was constructive of a new urban subject, one capable of blocking things out at will and developing what he called a "blasé attitude."[22] This metropolitan man, as he described it, was rational and calculating. To accommodate everyday life (talking to strangers, buying food, or commuting to work) he had to exercise a kind of mental reserve.

The city was incomprehensible in its unfiltered form, so having mental reserve was required to parse out the various social situations from the sights and sounds of the urban street. The blasé attitude was a coping device that people adopted to deal with the realities of urban life. Although Simmel was highly critical of this new urban subjectivity, he acknowledged the unprecedented freedom enjoyed by the metropolitan man. Life in the metropolis forced a seemingly unnatural rationalization of everyday life, but at the same time it enabled a freedom to become something different and break the bonds of small town life.

Fast-forward one hundred years, and Simmel's observations still resonate. At the time of Simmel's writing, he noted that the city, and its associated technologies, forced individuals to mentally adapt to its form. Urban subjectivity was a new way of approaching the world. Anonymity and increased sensory stimulation became a regular part of everyday life. And he recognized that there was no turning back. He made it clear that the metropolitan man can never return to the small town, as he would feel too restricted. The smaller the social circle, "the more anxiously the circle guards the achievements, the conduct of life, and the outlook of the individual."[23] In the metropolis, the individual reigned supreme. The personal freedoms enjoyed in the metropolis were seductive—even if they required a complex assortment of new technologies to realize them.

There is no doubt that the blasé attitude still exists—people place limits on what they take in through their senses. They might consciously choose to avoid looking at a homeless man or the electronic billboards that surround them, and they employ new technologies to assist in that filtering. Mobile devices are useful interfaces through which to engage the seemingly incomprehensible world around them. As more of the world's information is available online, it is possible for people to outsource some of that filtering to their mobile and location-aware devices.

As people employ technologies to assist in filtering, they produce more information to be filtered. They even reorient the nature of the space. Urban space extends into the web and vice versa. As Eric Gordon argues elsewhere, new technologies always prompt a change in urban subjectivity and urban form, and not necessarily to the exclusion of what came before it.[24] Urban spaces are constantly changing to accommodate new tools and practices. For example, the introduction of photography altered how people looked at cities, and so, too, did the popularization of film. Today, the rapid increase and accessibility of web data marks another important shift in urban spaces. Net localities have proliferated because of technologies that enable urban practices to extend beyond what one can touch or see. Of course, there are consequences to each technological shift and there is a period of social realignment where social norms adjust to the realities of everyday life.

Barry Wellman has pointed to a change in social interactions that he calls "networked individualism" where people connect directly to one-another through the network, but not to the physical space surrounding them.[25] Others have noticed trends of "telecocooning"[26] or "selective sociality,"[27] where small groups of friends interact via mobile digital networks and ignore the larger public sphere. The web has introduced new platforms for interaction and those platforms are almost certainly altering interactions in public spaces. But just as we can characterize this trend as a propagation of the worst elements of online life, where human connections are rationalized into computer code and public isolation, net locality can also be understood as yet another shift in the meaning of urban spaces and the freedoms (perceived or otherwise) associated with them. Traditional urban public space is perhaps becoming like the small town, where pure physically co-present social circles seem oppressively small. Not being connected to a network, not having access to location-based information, is tantamount to being closed off to a space's potential. Along the lines of Simmel's metropolis,[28] net localities encourage a new social organization that produces self-protectionism; but, at the same time, it produces a sense of freedom. The person participating in a net locality is not limited to what immediately surrounds her; she has the ability to associate with a much wider swath of information and people. Even Simmel recognized that the metropolis expanded the who and what of social interaction. Although people are still dependent on social groups, he says, "It has become a matter of choice with whom one affiliates and upon whom one is dependent."[29] Before net locality, the city made it possible to self-select people and spaces with which to associate. The rise of net localities makes this kind of selectivity the prominent characteristic of city space. For example, a college student with interest in math and hip-hop who is using a location-based social network like *LooptMix* may set his application to give alerts when other students who like math and hip-hop are around. But that might cause him to bypass other college students who enjoy classical music and study history. Likewise, locations to which digital information are "attached" might be more visible for location-aware phone users than other locations that don't show up on a mobile map.

But as net localities become the norm, they are not necessarily normalized; they still might appear disruptive or strange in the larger social context. From the outside, they might appear to be dead spaces, collections of co-located individuals with nothing to say to one another. But, in fact, there is a great deal of nuance in net localities—a great deal of social exchange that is slowly carving out new social rules and new ways of interacting with public spaces. To understand these new dynamics of social and spatial interaction, we need to look at specific practices of socialization. How are people using technologies to engage with the proximate and the remote simultaneously? What rules of conduct are assumed in these engagements and what new rules are emerging that expand the possibilities of how net localities are experienced?

HERE'S THE SITUATION

We turn to sociologist Erving Goffman[30] to clarify the actual practice of these minor social contracts that define net localities. This might seem like a counter intuitive approach, as Goffman was a theorist of face-to-face interactions, concerned primarily with a unit of analysis he calls the "situation," or "the full spatial environment anywhere." But while Goffman pays little attention to mediated interactions, he provides a very productive framework from which to approach the problem. There are social rituals that compose every situation; these rituals are organized by a set of expectations and the performances born of the meeting of those expectations.[31]

Goffman admits that society has an expectation that urban spaces be comprised of each individual's undivided attention. But he is quick to point out that engagement, even in important matters like urban space, is highly variable, and typically consists of two kinds of involvement: dominating and subordinate. "A dominating involvement," according to Goffman, "is one whose claims upon an individual the social occasion obliges him to be ready to recognize; a subordinate involvement is one he is allowed to sustain only to the degree, and during the time, that his attention is patently not required by the involvement that dominates him."[32] In urban spaces, the situation manufactured by the physical composition of things and people is dominating in most cases. When we are standing right next to others, even in non-intimate contexts like a public street, we expect the dominating involvement to be the physical space and the nearby person. However, in net localities, we acknowledge that there are other outlets for involvement, and we accept them, insofar as they don't interfere with the established order of the dominating involvement.

When a person sending a text message while walking bumps into somebody else, this is an obvious affront to social expectations. When a person moves to the side of a crowded sidewalk to send that text message, the correspondence between the dominating and subordinate involvements is not as clear. That person is ostensibly removing themselves from their dominating involvement with the physical space, but they have not caused any harm or inflicted any injury. They are, however, seemingly ignoring their obligation to participate in the outward appearances of a space that feels truly public. People do this all the time, with or without the aid of mobile devices. Reading on a train removes a person from the dominating involvement of the space. Walking around with an iPod functions the same way. Even daydreaming on a street corner can momentarily "take someone away." But there is a middle ground created when a person "goes away," but does so while maintaining the dominating involvement of the local space. Goffman recognizes that when people look as *though* they are coming from someplace or going to someplace, they exhibit an "objective that leaves the actual focus of attention free for other things; one's destination, and therefore one's dominant involvement, lie outside the situation."[33] In this case, there is a disconnect between the typically dominant involvement of the physical space and the person's focus of attention which rests outside of it. But what happens when the person's attention is focused on the map of where they happen to be standing, or a person in their social network that happens to be down the street? The dominating involvement is not limited to the physical situation—it is clear that the physical situation remains integral to the larger situation. Looking at nearby restaurant reviews on an iPhone map momentarily distract one's attention, but with the goal of applying that distraction to the physical situation. While it might appear that the person is "lolling," as Goffman calls it, they are, in fact, deliberately extending the purview of the local situation.

But perhaps it does not matter. Looking down at a device in a situation where other co-present individuals expect that you exhibit attention to the physical space might appear disruptive, regardless of individual intent. The user, then, is often responsible for maintaining two separate situations, each with a set of rules that need to be followed. In public space, there is a rule against "'having no purpose,' or being disengaged."[34] As a result, users appropriate "untaxing involvements to rationalize or mask desired lolling—a way of covering one's physical presence in a situation with a veneer of acceptable visible activity."[35] This might mean sitting on a bench, or stopping in a nearby alcove with one's mobile phone. In the context of net localities, there are rules as well. There is etiquette for stopping what you are doing and checking in to a location-based social network, like *Foursquare* or *Loopt*, and there are prohibitions against ignoring someone in your network that is "very close." If a known friend is one block away and just checked in to *Foursquare*, she might not be physically co-present, but it would be rather rude to ignore her. This situation would require that the user cover up their involvement by appearing "otherwise engaged." Perhaps an update such as "I'm in a meeting," or, "Rushing to get the train." In net localities, the local space is the dominating involvement; however, the local space is not always solely physical—it also includes digital information and networked connections.

Within net localities, therefore, the coherence of the physical situation remains, while the user's attention is freed up to an ecology of foci which, rather than destroying urban spaces, are constructive of people's experiences in these spaces. That person on the street might be sending a text message to a colleague halfway across the world, but they also might be searching for information about where they are standing, in the form of restaurant reviews or urban history, or communicating with a person who happens to be in the coffee shop across the street. The composition of net localities can tolerate a wide variety of attentional foci as long as the physical situation is not disregarded (as in the case of the texter running into somebody else). In fact, as the accepted norms of urban spaces shift to accommodate networked interactions, social expectations and rules of etiquette will shift along with them. In net localities, the situation is understood to be larger than what is physically near.

"GETTING AWAY WITH GOING AWAY"

While Simmel lamented that the metropolis forced individuals, as a condition of their freedom, to shield themselves with a blasé attitude—a way of (dis)engaging with the world characterized by a rational and calculating reserve, Goffman approached the problem of continued and intense involvement in urban public spaces from a different, yet complementary angle. Instead of assuming permanent changes in a mental state to deal with the consistently chaotic state of the metropolis, Goffman noticed that, in public social situations, people grasp on to opportunities to momentarily escape. He called this "going away:"

> When outwardly participating in a social activity within a social situation, an individual can allow his attention to turn from what he and everyone else considers the real or serious world, and give himself up for a time to a playlike world in which he alone participates. This kind of inward emigration from the gathering may be called "away."[36]

Away, for Goffman, meant a mental retreat into another space. It meant not being in the situation. It didn't, however, necessarily imply a disruption to the situation. Individuals can "get away" while the social ritual that organizes the situation remains intact. For instance, letting one's mind wander on a crowded sidewalk is often aided by looking at a billboard, a magazine, or a mobile phone. Goffman called this "getting away with going away."[37] We focus attention on things in the physical environment to create the impression that we are only momentarily distracted from the situation.

Or, increasingly, it is not uncommon for people to begin from the state of "away" so that they have control over when they return. Take, for example, the use of earbuds and an mp3 player on a crowded street corner. If one runs into a casual acquaintance while waiting to cross a street, and that acquaintance has buds in her ears, she has permission to behave somewhat differently in the situation. She can smile and nod, while keeping the earbuds in place, and then turn her head to once again go away. Or, if she really wanted to talk to the interlocutor, she could remove her earbuds and engage in the kind of small talk that would normally be required for this casual connection. If, on the other hand, she did not have the earbuds at all, and wanted not to be insulting, she would have to go with the latter option. Social regulations would restrict her from turning away and ignoring the encounter all together. In this case, the technology provides a means of filtering the city, or creating an outward appearance of the reserve to which

Simmel referred, but in such a way that empowers the individual to engage in the situation on her own terms.

In this example, the individual starts from the position of away, and instead of getting away with going away, she is getting away with *being* away. And her dominating involvement is likely someplace other than the local situation. This example demonstrates how wielding control over the terms of engagement is important for an individual's sense of mastery over a space, but may not be conducive for "good public spaces." Net localities, on the other hand, are places in which individuals wield control of information, while keeping their dominating involvement local. Even as they "go away" from the physical situation to review messages on their iPhone, they are using the local to organize their involvement in the digital interface. While there are still social regulations that organize physical and digital spaces independently of one other, because the dominant involvement is the same, there is a relaxation on the level of policing that each situation receives. If three people are walking down a street looking for a restaurant and one of them "goes away" to look at the *Yelp* iPhone application that finds nearby restaurant reviews, it is easy for him to "get away with going away," as his leaving the physical situation is understood as being in service to the net locality. As Goffman explains:

> In public, we are allowed to become fairly deeply involved in talk with others we are with, providing this does not lead us to block traffic or intrude on the sound preserve of others; presumably our capacity to share talk with one another implies we are able to share it with others who see us talking. So, too, we can conduct a conversation aloud over an unboothed street phone while either turning our back to the flow of pedestrian traffic or watching it in an abstracted way, with the words being thought improper; for even though our co-participation is not visually present, a natural one can be taken to exist, and an accounting is available as to where, cognitively speaking, we have gone, and, moreover, that this 'where' is a familiar place to which we could be duly recalled should events warrant."[38]

As Rich Ling points out, this is one of the few passages in Goffman's writing where he actually refers to mediated conversation.[39] What's important in this example is not what is on the other end of that unboothed street phone, but that the person talking on the phone has gone away somewhere familiar. As long as the other people in the situation understand where he has gone, the act of going away is less likely to disrupt the situation. In a net locality, if one goes away as a means of enhancing local interaction and convincingly demonstrates the continuity in their dominating involvement, then it is quite easy to "get away with going away" without sacrificing the internal or external qualities of the shared place.

PERFORMANCE IN/OF PUBLIC

Information is not just something to consume. One's awareness of nearby information (and people) can also be a context for performance. When urbanist William Whyte described "good public space," he referred to the Seagram Building plaza in Manhattan, where "on a good day, there would be a hundred and fifty people sitting, sunbathing, picnicking, and schmoozing—idly gossiping, talking 'nothing talk.'"[40] For Whyte, the publicness of the space was the outward appearance of people engaging with one another, even if those engagements had no content ("nothing talk"). Architects and urban designers have so completely bought into this representation of public that, for decades, they have tried vigorously to reproduce similar spaces, if not in actual fact than only in appearance, by designing pedestrian intersections and street furniture to produce the appropriate impressions.[41] This conceptualization of public has a long history in the digital realm as well. Since the 1990s, designers of chat rooms, MUDS, and MOOs have represented spaces that gave the impression of being public.[42] In LambdaMOO, for example, private spaces were referred to as "bedrooms," and public spaces as "living rooms." The intention was to create spaces for idle talk, for serendipity, where in the spirit of "good public spaces," people would just happen upon one another.

So, while good public spaces were spaces that invited idle chatter, great public spaces were spaces that mandated it. Whyte refers to the plazas in Rockefeller Center in midtown Manhattan as a great space. It is great because it maintains its publicity through performance. Despite popular opinion, he says:

> The lower plaza is only one part, and it is not where most of the people are. They are in the tiers of an amphitheater. The people in the lower plaza provide the show. In winter, there is skating; in summer, an open-air café and frequent concerts. The great bulk of the people—usually about 80 percent—are up above: at the railings along the street, along the mezzanine level just below, or on the broad walkway heading down from Fifth Avenue.[43]

While praising the space, he laments that it is often misunderstood. Architects are constantly borrowing from the design of the plaza, but as means of creating a space for crowds, they typically reproduce only the lower plaza, without its context. "They wind up having a stage without a theater, a hole without the doughnut. And they wonder what went wrong."[44] Watching a performance (or performing), whether official or unofficial, is part of what makes public spaces work. Watching the ice skaters from Fifth Avenue is an acceptable form of going away. And, indeed, ice skating in Rockefeller Center, as a kind of performance, is also a kind of going away. In the ideal situation, the occupant of the space can move between these two practices near seamlessly.

Goffman acknowledges the importance of performance in everyday interactions. He goes so far as to use the analogies of a stage with its front and back regions.[45] When on stage, people behave in a manner dictated by the rules and regulations of a performance. Actors remain loyal to the script and the audience expects that they do so. But they do not have to always remain on stage; they may retreat backstage where social regulations are not quite as policed. After a scene, an actor may leave the stage and complain to the stagehands that the "audience seems really dead tonight." And yet, when he returns onstage, he is expected to conceal this opinion. In one example, Goffman describes a waiter. The waiter behaves differently when in the dining room (front-region) than when in the kitchen (back-region). He speaks to the patrons of the restaurant in one way and he speaks to the other waiters and kitchen staff in quite another. What is interesting about this situation is that all the actors understand the distinction between the front region and back region. The restaurant patrons know that the waiter might retreat backstage and speak in a different manner; and of course the kitchen staff knows that the waiter will have to perform when he goes "onstage." The performance remains stable just as long as the regions do not blur. If the kitchen conversation were to be overheard in the dining room, the performance would be disrupted. So consider again what makes Rockefeller Center work. It's not just that there are the two plazas: one for the performer and one for the audience. But it is a "great public space" because one doesn't have to remain on or off stage. There is fluidity between the two spaces—where performance can become voyeurism, and voyeurism can become just an appearance of voyeurism as one attends to other matters. And each act of performance and voyeurism references the other. The local situation is central.

This dynamic defines net localities. Consider the location-based social network, *Foursquare*, as an example. It is built on small performances. When a user checks in to a location with their location-aware mobile device, they are announcing their presence in that location. Metaphorically, this is like inhabiting the lower plaza while one's "friends" gather around to observe. While physically occupying a given location, the person checking in needs to perform to the mediated audience and as such needs to go away for the amount of time it requires to check in. Though one could argue that this practice primarily serves to disengage the performer from the situation, it is possible to say the same about the ice skater in Rockefeller center. The dominating interest of the ice skater is clear and observable; and the dominating interest of the *Foursquare* user is less clear. However, the outcomes are comparable. Whether standing on Fifth Avenue or using *Foursquare*, the place of the performance becomes the spectacle for the observers and the place of the observers becomes the context from which the public space garners meaning. While the design of physical space matters for the organization of social situations, *Foursquare* demonstrates that information flows can be equally as influential. With this, Whyte's concern about the architect's lack of consideration for the variable spaces of performance is brought into relief. Just as people often misunderstand the functionality of the plazas at Rockefeller Center, so too do people misunderstand how information flows influence the possibilities for engaging in net localities. There is evidence that people are becoming more brazen with their use of technologies in urban spaces, choosing remote contacts over proximate connections.[46] But there is also evidence that going away does not require turning one's back on the local situation; it does not require that the user is operating outside of the context of the place he or she is inhabiting through networked extension. "Going away" is definitive of net localities; it is not simply a locality disrupted by a network. As the use of location-aware technologies continues to expand in urban places, the nature of the city will change.

CONCLUSION: TRANSFORMED URBAN PLACES

Simmel ends his 1901 essay "The Metropolis and Mental Life" suggesting that because "such forces of life have grown into the roots and into the crown of the whole of the historical life in which we, in our fleeting existence, as a cell, belong only as a part, *it is not our task either to accuse or to pardon, but only to understand*" (emphasis added).[47] Likewise, we seek to understand net localities so that we might shed light on how to design the technologies that mediate them, as well as to grasp what it is to inhabit these new types of public spaces and enable new kinds of places. People are using technologies in a variety of ways and on a regular basis, bringing the reality of network access to bear on everyday interactions in urban space and place. In some cases, they are using these tools to distance themselves from public space, "getting away with being away," but in other cases, they are using the tools to expand the purview of public space by bringing local information in networks to bear on local spaces. In net localities, what Aristotle would call the enclosure of physical space is expansive, not limited by the traditional boundaries of geographical extension. The material from which net localities are constructed include dynamic networked connections, remote and local social interactions, and location-based information. It is the *presence* of this expansive material and the emerging processes with which people engage it that constitutes net locality.

The pervasiveness of location-aware technologies in cities does not necessarily lead to the disintegration of "good urban spaces." Net localities provide a counter example for all of those who decry the effects of technology on public space. Public urban spaces are far too complex to diagnose. There are many uses for urban spaces and location aware technologies are opening up more of them. Good, vibrant public spaces will accommodate these tools and become a platform for the various modalities of constructing and communicating urban places.

NOTES

1. An extended version of this chapter is available in our book, Eric Gordon and Adriana de Souza e Silva, *Net Locality: Why Location Matters in a Networked World* (Boston: Blackwell Publishers, 2011).
2. Paul Goldberger, "Disconnected Urbanism: The Cellphone Has Changed Our Sense of Place More Than Faxes, Computers and E-Mail," *Metropolis Magazine*, February 22, 2003, http://www.metropolismag.com/story/20070222/disconnected-urbanism.
3. Howard Rheingold, *Smart Mobs: The Next Social Revolution* (Cambridge, MA: Perseus Publishing, 2002).
4. Rich Ling, *The Mobile Connection: The Cell Phone's Impact on Society* (San Francisco: Morgan Kaufman, 2004); Kenneth Gergen, "The Challenge of Absent Presence," in *Perpetual Contact: Mobile Communication, Private Talk, Public Performance*, eds. James Katz and Mark Aakhus (New York: Cambridge University Press, 2002); Keith Hampton and Neeti Gupta, "Community and Social Interaction in the Wireless City: Wi-Fi Use in Public and Semi-Public Spaces," *New Media and Society* 10, no. 6 (2008): 831–850.
5. Norman Klein, interview with de Souza e Silva (November 8, 2002).
6. Gergen, "The Challenge of Absent Presence," 227.
7. Shaun Moores, "The Doubling of Place: Electronic Media, Time-Space Arrangements and Social Relationships," in *Media/Space: Place, Scale and Culture in a Media Age*, ed. Nick Couldry and Anna McCarthy (London: Routledge, 2004).
8. Manuel Castells, "Materials for an Exploratory Theory of the Network Society," *British Journal of Sociology* 51, no. 1 (2000): 5–24.
9. Doreen Massey, "The Conceptualization of Place," in *A Place in the World? Places, Cultures and Globalization*, eds. Doreen Massey and Pat Jess (Oxford, UK: Oxford University Press, 1995), 45–85.
10. Joshua Meyrowitz, *No Sense of Place: The Impact of Electronic Media on Social Behavior* (New York: Oxford University Press, 1985).
11. Doreen Massey, "Power-Geometry and a Progressive Sense of Place," in *Mapping the Futures: Local Cultures, Global Change*, eds. Jon Bird, et al. (London: Routledge, 1993), 66.
12. Henri Lefebvre, *The Production of Space* (Malden, MA: Blackwell Publishers, 1991).
13. David Harvey, "From Space to Place and Back Again," in *Mapping the Futures: Local Cultures, Global Change*, eds. Jon Bird, et al. (London: Routledge, 1993), 23.
14. Steve Harrison and Paul Dourish, "Re-Place-Ing Space: The Roles of Place and Space in Collaborative Systems," in *Proceedings of the 1996 ACM conference on Computer supported cooperative work* (Boston: ACM, 1996), 69.
15. Yi-Fu Tuan, *Space and Place: The Perspective of Experience* (Minneapolis, MN: University of Minnesota Press, 1977), 73.
16. Gaston Bachelard, *The Poetics of Space* (Boston: Beacon, 1994), 6.
17. Adriana de Souza e Silva and Daniel M. Sutko, "Theorizing Locative Technologies through Philosophies of the Virtual," *Communication Theory* 21, no. 1 (2011): 23–42.
18. Licoppe and Inada call this an onscreen encounter. See Christian Licoppe and Yoriko Inada, "Emergent Uses of a Multiplayer Location-Aware Mobile Game: The Interactional Consequences of Mediated Encounters," *Mobilities* 1, no. 1 (2006): 39–61. What's interesting about these encounters is that they don't exist in opposition or even in parallel to physical

co-presence, but increasingly, they are experienced in dialogue with the physical situation.

19. Adriana de Souza e Silva, "From Cyber to Hybrid: Mobile Technologies as Interfaces of Hybrid Spaces," *Space and Culture* 3 (2006): 261–278.
20. Georg Simmel, "The Metropolis and Mental Life," in *On Individuality and Social Forms; Selected Writings* (Chicago, IL: University of Chicago Press, 1971); and his *The Sociology of Georg Simmel*, trans. K. Wolff (New York: Free Press, 1950).
21. Erving Goffman, *The Presentation of Self in Everyday Life*, Rev ed. (New York: Doubleday, 1990); and his *Behavior in Public Places: Notes on the Social Organization of Gatherings* (New York: Free Press of Glencoe, 1963).
22. Simmel, "The Metropolis and Mental Life."
23. Simmel, "The Metropolis and Mental Life," 330.
24. Eric Gordon, *The Urban Spectator: American Concept-Cities from Kodak to Google* (Hanover, NH: Dartmouth College Press, 2010).
25. Barry Wellman, "Little Boxes, Globalization, and Networked Individualism," in *Digital Cities II: Computational and Sociological Approaches*, ed. Makoto Tanabe, Peter van den Besselaar, and Toru Ishida (Berlin: Springer, 2002).
26. Ichiyo Habuchi, "Accelerating Reflexivity," in *Personal, Portable, Pedestrian: Mobile Phones in Japanese Life*, eds. Mizuko Ito, Daisuke Okabe, and Misa Matsuda (Cambridge, MA: MIT Press, 2005).
27. Misa Matsuda, "Mobile Communication and Selective Sociality," in *Personal, Portable, Pedestrian: Mobile Phones in Japanese Life*, eds. M. Ito, D. Okabe, and M. Matsuda (Cambridge: MIT Press, 2005), 123–142.
28. Simmel, "The Metropolis and Mental Life."
29. Simmel, "The Metropolis and Mental Life," 130.
30. Goffman, *Behavior in Public Places*.
31. For more on Goffman and mobile phones, see Rich Ling, *New Tech New Ties: How Mobile Communication is Reshaping Social Cohesion* (Cambridge, MA: MIT Press, 2008).
32. Goffman, *Behavior in Public Places*, 44.
33. Goffman, *Behaviour in Public Places*, 56.
34. Goffman, *Behaviour in Public Places*, 58.
35. Goffman, *Behaviour in Public Places*.
36. Goffman, *Behaviour in Public Places*, 69.
37. Goffman, *Behaviour in Public Places*, 70.
38. Michael Goldhaber, "The Value of Openness in an Attention Economy," *First Monday* 11, no. 6 (2006): 86.
39. Ling, *New Tech New Ties*.
40. William Whyte, *The Social Life of Small Urban Spaces* (Washington, DC: Conservation Foundation, 1980).
41. Michael Leccese, Kathleen McCormick, and Congress for the New Urbanism, *Charter of the New Urbanism* (New York: McGraw Hill, 2000).
42. Julian Dibbel, *My Tiny Life: Crime and Passion in a Virtual World* (New York: Henry Holt and Company Inc., 1999); Howard Rheingold, *The Virtual Community: Homesteading on the Electronic Frontier* (Cambridge, MA: MIT Press, 1993).
43. Whyte, *The Social Life of Small Urban Spaces*, 59.
44. Whyte, *Social Life*, 59.
45. Erving Goffman, *The Presentation of Self in Everyday Life* (New York: Anchor, 1959).

46. Chantal De Gournay, "Pretense of Intimacy in France," in *Perpetual Contact: Mobile Communication, Private Talk, Public Performance*, eds. James Katz and Mark Aakhus (Cambridge, MA: Cambridge University Press, 2002); Habuchi, "Accelerating Reflexivity."
47. Simmel, "The Metropolis and Mental Life," 340.

6 The Real Estate of the Trained-Up Self
(Or is this England?)

Caroline Bassett

[S]harpen your mind wherever you are . . . Speed Brain is a healthy addition to your iPhone! (Lumosity.com website)

Whose streets? Our streets! (adopted by UK students demonstrating against university fee increases)[1]

In a land where neoliberalism is actively terminating a certain kind of shared knowledge project (notably, the arts and humanities), where facts and skills are increasingly viewed as personal capital to be taken to market and "knowledge" is viewed as an object to be purchased, brain-training games (e.g., those found on Lumosity.com[2]) that are sold as painless, anytime, anywhere, personal aerobics for the brain—as mobile cognitive gymnasiums—seem a disturbingly tight fit. These games promise cognitive augmentation, rising efficiency, better concentration, and thence gains in personal productivity. Regular play using a suite of games accessible from a series of platforms promises to make you smarter. Your "Brain Processing Index" will rise and you will perform better, both absolutely and in relation to other players, and not only in the world of the game but also in the game of life—to deploy an increasingly empty but nonetheless loaded phrase, these games develop eminently "transferable skills."

Lumosity.com's promises are typical of contemporary brain-training products. The packages on offer bundle rigor (science, structure, the program), with ease (rapid improvement through exercises that take only a few minutes a day) and flexibility (ad hoc, anytime, anywhere use). Many earlier and more puritan forms of training shared the same instrumental goal of self-improvement; however, the methods were often rather different. Charles Dickens' utilitarian schoolteacher Mr. Gradgrind, for instance, famously dealt in content (the acquisition of "facts"),[3] whereas these games eschew content and instead claim to sharpen the senses and train cognitive facilities such as memory, mental dexterity, concentration, and attention. Moreover, in contrast to Gradgrind's absolute division between work and play (and their fixed locations in the classroom and the circus, respectively), in the case of

Lumosity.com, these are entwined: These games are sold as serious forms of fun (promising real-world results) and as fun forms of being serious (results come through enjoyable play). Designated as an all-ages mode of edutainment, they are not designed for timetabled play in the classroom, but for use in everyday spaces and places, including those that are conventionally designated public. What I am interested in teasing out here are the questions about place (and the relationship between various dimensions of contemporary place) that are raised by playing these self-improvement games in public.

Brain-training games line up with a contemporary preoccupation with self-improvement, an aspect of the more general project of self (variously explored by Giddens and differently by Foucault) that is said to be a central trope of modernity that is at once intensely personal but also social—a dialectic that can be expressed in spatial terms: gamers are subjects in space, and gaming is a practice in space that contributes to the making of place and to the making of relations of place.

Today's everyday urban places are complex, over-layered, "multiplex"[4] ecologies, combining fully proximate elements (human and non-human) with less substantial inputs from the more remote sites (communicational traffic of various kinds) and incorporating through that layering various temporalities. The micro activity of gaming—multiplied many times over, and each time enacting and also endorsing the enactment of a particular practice of space—contributes, if in small ways, to the production of a space as a place with particular characteristics. It thus adds a new ingredient to the ecology, tilting the balance of the multilayered spatial and temporal dynamics of a specific place, for instance the degree to which it may operate in more or less connected, accessible, social ways. Micro-gaming thus contributes to a more general reformation of the moral economy pertaining in relation to public places—to a change in ideas about how such space should be "rightly" used.

It is useful to begin by questioning the common assumption that density of digital devices produces density of connection. This neglects the specificity of devices (even converged devices), a specificity that emerges as they are put to use (because it is in use that meaningful distinctions may be made—between what technical architectures enable and what socio-technical formations produce and between what games model and what they do and make as they are practiced). Exploring mobile phones, Vince Miller argues that the rise in phatic culture in contemporary web sociality is evidence of the valorization of connected presence, supported by pervasive networks and intimate devices.[5] He connects this "virtually compulsive" desire for connectivity to the broader social contexts of what Beck and Giddens have discussed in terms of individualization and the project of the self.[6] In contrast, I argue here that the form of gaming I am exploring, which given its emphasis on individualization as a social priority also reflects to this broader formation, does something different. That is, it complicates as much as confirms the desire for, and production of, connectivity—where the latter is understood as entertaining a mode of local, place-based inter-action that might be understood as convivial.

In tension with digital media's tendency to produce connection (and the latter is the dominant expectation of pervasive computing) are the various forms of distancing, disconnection, and separation it also provokes. And if intimate media in Miller's account produces a desire for personal con-nection, this may be at some distance from, may conflict with, what I am tempted to call "common" connection, particularly common connection *in-place*. The vision of the everyday environment of the pervasively net-worked city as an increasingly public place needs to be tempered. This is explored here not only as a tension between connection and disconnection, but as a question of place and its dimensions, and as a tension between the formal sequestration of the game and the ambiguous location of play.

SCALE AND THE IMPORTANCE OF BEING TRIVIAL?

Underpinning this article is a personal ethnography of small game use in public (see my earlier work).[7] The attempt was to record an experience of gaming and to ask how gaming placed me in relation to the tissues and skeins of permanent and ephemeral elements—people, habits, objects, archi-tectures—that constituted a local environment. The scale of the attempt was small—appropriately so, because brain-training games themselves are short and fast. They are everyday bits and pieces, taking up small frag-ments of space and time, and they produce a glancing, if intense, engage-ment with under-elaborated screen environments. Their casual portability, short duration, and visual simplicity sets them in stark contrast to highly elaborated networked virtual environments offering sustained engagement in virtual universes for players—and sustained sites for various forms of academic study and dispute, notably around the rival claims of narrativ-ity and the ludic.[8] Instead, these games, this kind of gaming, this kind of micro approach, might fall within the purview of what the French theorist Georges Perec called the "infra-ordinary" and defined as a scale, something indicating both a form of life, and a suitable method to investigate it.[9] The nascent discipline of game studies rather lacks a sense of the trivial—and sometimes distinctly wants to emphasize its scientific approach.[10] Games scholars variously seek to establish the value of games' aesthetic qualities,[11] make claims for significance based on a valorization of popular over high culture or scarcity over mainstreaming—"cultural marginality warrants . . . truth value"[12] or focus on the market potential the games industry. These arguments have their merits. However, given the two-pronged pro-motional appeal, to "science" and to the discourse of self-improvement, that the training games make, it seems useful to go with Perec and cleave to the essential triviality of these games—trying to find a way of approaching them that registers their lack of importance and also explores what they might reveal *because* of their trivial scale—for instance, what they might suggest about intimate dispositions of bodily interaction that might not easily be got at any other way. Perec's approach to the exploration of the everyday was endotic. He sought to find ways to turn inward, to inquire closely into the minutiae of experience—often his own experience—and to do so in ways that were only barely methodological. As he put it:

> It matters little to me that these questions should be fragmentary, barely indicative of a method, at most of a project. It matters a lot to me that they should seem trivial and futile: that's exactly what makes them just as essential, if not more so, as all the other questions by which we have tried in vain to lay hold on our truth.[13]

In the following sections, the questions I have asked about these games are opened up further, first, through a consideration of play as rehearsal and/or performance, and, second, in relation to artificial memory and the dimensions of interior place. Third, I consider brain games as self-renovation projects that promise to reduce personal lag. Finally, I ask what these training games might seek to teach us about public places and how to use them.

EVERYDAY LIFE: PRACTICE, REHEARSAL, PLAY

Brain-training games set you up in your own private cognitive gym, providing a collapsible/inflatable space made of the same digital material that has, over the years, enabled many forms of collaboration, from temporary autonomous zones[14] to crowd sourcing projects.[15] In this case, however, the zone of directed interaction the screen delineates is personal; the equipment is a series of activities—memory games, numbers exercises, word puzzles—designed for your sole use. None last more than a few minutes, some are tedious, some oddly engaging; most are somehow both—flow states emerge in surprising places, exercises training peripheral vision connect powerfully with memories of real activities.[16] This, of course, is a personal judgment, and some of these states are only available by refusing to play the game "properly," but it serves to raise some questions about the rewards found in this kind of play, in particular about their location.

Adventure gaming offers a space that mimics extreme hazard; it is enjoyable, ultimately, because it is danger-free, having no consequences beyond itself. The "game over" declaration releases the player even while terminating the game-world character. Clear rewards are found in-game—and thrills heightened by modes of interactive identification close-coupling the player to the internal world of the game don't change that (though they may produce what Juul calls the half-real sensation of gaming that is part of its attraction). By contrast, puzzle games rely on the hazarding of competency and the satisfaction of mastery, rather than exposure to fictional danger. They, too, however, offer rewards in their own terms: success in the game results in a high score. Brain-training games do consist of puzzles. However, finding solutions is not really the point of these games because, although they demand fierce levels of on-screen attention, it is boosting performance in another place that is the real issue here. Alongside the collateral satisfactions of in-game mastery that the puzzles provide,[17] and/or the pleasures of a flow state, brain trainers are told they can expect to find the real rewards for their cognitive-muscle-building regime of hard labor in the game of real life. Cognitive work in-game produces enhancements pertinent to cognitive operations within everyday life. For instance, concentration and memory skills built in-game are to be transferred to other zones of everyday life.

Not that the two are divided. The body, playing in-game and simultaneously operating in places beyond the game, is a hinge here. It both defines (through attention) and denies (through embodiment) that unstable frame that divides and connects different zones of activity (different dimensions of place) and that has caused so much of the "trouble with the virtual" in new media studies.[18] We might say that this body, operating in two dimensions at once, at once performs in a space and rehearses in it; but with the caveat that both the performance and the rehearsal take various forms of place or are complicit in particular forms of emplacement.

Anthropologically inflected versions of media studies (see, e.g., Roger Silverstone)[19] brought together various theories of identity performance and critical cultural geography in interesting, if sometimes problematic, ways— notably weaving together Erving Goffman's analysis of performance and identity, based on symbolic interactionism, and the thinking of Henri Lefebvre and the French critical tradition, with which Perec is loosely associated.[20] It is useful to draw on these traditions but also to disentangle them somewhat: each suggests ways to consider distinctions between game play as performance and rehearsal, two relatively distinct, although also overlapping, ways of practicing space.

The focus on performance points to Goffman's explorations of the production of the self, a process accomplished on the various "stages" and sites of everyday life; sites that are often, in his research, the more or less "sequestrated" places of institutions (e.g., prisons, hospitals, and educational institutions).[21] Goffman's focus is on small worlds and tiny actions, but his work on the interaction between the performer, audience, and stage underscores the degree to which identity performance is, despite its intimate scale, always interpersonal rather than personal, "interactional" rather than "idiosyncratic,"[22] and also situated. Moreover, Goffman, stressing interaction as the basis of the production of the self, queers the division between public and private space. In doing so, he complicates and even confounds—as feminism itself did (e.g., in the claim that the personal is political)—the division between public and private in these games of life. (This is one reason why, as Candace West convincingly argues, feminism has more of a debt to Goffman than it may recognize.)[23] Elsewhere in his work, Goffman develops these questions of personal performance and its necessary production in public (on the public stage), in relation to interpersonal relations, through his analysis of modes of "civil inattention:"[24] ways of not looking at, or of overlooking, others encountered in a public zone, no longer policed by its division from a private realm. This might be used to explore the dynamics of spaces simultaneously used as rehearsal (back stage) and front stage (performance) space, to ask questions about a public economy of attention, in which non-symmetrical relations of presence and absence are imagined and willed as well as materially instantiated.

If West takes from Goffman a certain porousness between categories of public and private and the performance of the self within them, Mark Selzer, exploring Goffman's work in relation to parlor gaming, focuses on sequestration as a model both of the social world and as a model of gaming. This allows him to argue that the *relation* between model worlds (tiny worlds) and the world as model is essentially fractal.[25] The conception of the game as a closed world (even a sequestrated space?) with its own dynamics is familiar. Huzinga, for instance, long ago argued that play is a "stepping out of real life into a temporary sphere of activity with a disposition all of its own."[26] Selzer moves this argument on by arguing that the games he explores are models of "real life" because they are models. In this context, he argues that the sequestrated worlds of games (in his case, directly antagonistic parlor games) are dress rehearsals for real life (in his case, for the "military entertainment complex"[27]), not only in the sense that they may build particular real-world skills (e.g., Margaret Tan[28]), partly by exploiting a discrete physics that, for instance, translates risk/damage into an abstraction, but also because they at once model and realize a "social territory."[29]

Let me now turn back to the games I am concerned with. It might be said that sequestration—in mobile circumstances—is typically partial, partly virtual, a question of separated and shared layers, a matter of augmentation, or overlays, and a matter of dimension rather than distance. In the place of the parlor and parlor game—but also in the place of the training academy of Gradgrind and his unfortunates, frightening in its isolation and intensity—is the training game undertaken in the semi-transparent and overlaid space of the pop-up cognitive gym and the real-world space within which it is enmeshed. The game space is molecular rather than disciplinary if you want to see it that way.[30] Even the unity of space characteristic of the parlor game—with its real players (playing with unreal objects) who, through their physical presence as antagonists, contribute to its half-real quality, according to Selzer (who is drawing on Juul)—is fragmented.[31]

To train in public is not to be involved in the same kind of dress rehearsal as Selzer finds in parlor gaming, not least because, although it does involve rehearsal (work on cognitive skills to improve mental performance), it also entails performance now (play in a public place). In the case of Lumosity.com, the act of "finding time" and "using space" to play the games *itself* marks a shift toward a more efficient way of "doing time" and "doing space." It might be said that the place I am in is not unified through play, thus becoming, in Selzer's words, a small world, but is instead divided—or, from my perspective as the player, it is re-doubled. The take-away lessons from Lumosity.com are not only centered on cognitive improvement (brain power) but efficient social interaction, here understood as efficient use of shared place. The Lazarus principle, beloved by digital industry commentators, says that previously dead time can be revived and given back to individuals through mobile information and communications technologies (ICTs).[32] These brain-training games are given to users as a means by which the dead time of odd moments—here the "wasted" spaces of ambient collectivity or sociality—can be used for personal efficiency gains. Lumosity.com tells users that they must play now to train for future performance (users play the game to improve cognitive efficiency), but, by playing, users are also/already improving personal efficiency at the time of play by making efficient use of what might be seen as "wasted" time or "wasted" space.[33]

Lumosity.com thus enacts a lesson in "how to use time" and "how to use space," particularly public space. Playing these games anytime, anywhere, a user learns to use public space as her own personal training room. And this space may not even be viewed as a "half-live" space: because, insofar as the user makes this larger public space—in all its dimensions—function only as a game space, she does not share it with real partners or antagonists. Marc Augé[34] noted that place and non-place were in part a matter of emotional attention, essentially arguing that "through an intense experience (or practice) of these spaces by its practitioners, they can be turned into places."[35] This proposition can be reversed; when enough directed intensity is removed from a certain dimension of space, it might become, partially at least, a non-place—even while it is being *used*. This can also be explicitly explored in relation to social interaction: a public economy of attention and interaction. Or, perhaps we could say that this re-doubled inattention leaves a hole in the fabric of the public, a dead end—in terms of communication and circulation, if not consumption.

PLACE AND THE DIMENSION OF MEMORY

The fractal sense of the world as a series of model spaces, games in which play becomes a rehearsal of action to be undertaken again in other models, stands in sharp contrast to the conception of play bound into various critical/situationist interventions into everyday life. Play, and the playing of games as a mode of inquiry ("not even a method"), was an important tool for the situationist fraction of the everyday life theorists. As Lefebvre famously put it, everyday life is both the site of the worst and the best, the site of something authentic as well as the site for the most nearly experienced alienation.[36] Its promise is that it can be the site where activities might be undertaken to reveal or produce *specificity*, to break the equivalence model of commodity fetishism (relations between people disguised as indifferent relationships between things), to break the given rules of the game. Georges Perec's injunction to attend to the infra-ordinary by listening in to the self resonates with this, being based in part on a sense, shared with Lefebvre, that the ordinariness of everyday life conceals the marvelous[37] possibilities it holds and that these possibilities can flourish. Perec's work refuses the systematization of Lefebvre, and the model of space (the production of space) as a tri-partite production, that the latter provides. As we have already seen, Perec's method of investigation is playful and partial, and although it makes use of systems such as numerology, small rules, database sorts, and oblique tactics (such as, e.g., the careful inspection of specific classes of objects, the making of obscure lists, and the generation of tax-onomies through various forms of numerology), it always seeks to move beyond rule-given "results."

It is through a form of play that Perec seeks to enter, not the seques-trated institutions modeling everyday life, but the sequestrated aspects of the self—the obscured treasure that is one's own set-aside memories. Thus, sequestration itself might be frustrated through a mode of playful archeol-ogy, one in which the subject is operated on. Where Selzer's game models tend to confirm the pre-given rules of the larger game (which also produces a focus on the anterior),[38] Perec's work has the disruptive effect of play-ing games on the game of life itself. He does this above all in relation to memory and often in relation to memory places.

Many of the Lumosity.com games train various forms of visual and spatial memory.[39] Once again, these games demand concentration and train attention. The tasks are to be completed against the noise of con-flicting signals—sound contradicts vision, written information interrupts color, edge movement distracts from central vision tasks. This produces a demand to block out the exterior world—look up and you are done for, but these games also train against the habit of looking "in." Journeys into the imagination, returns to memorized places, do not help play this game, but rather constitute a slip of attention and guarantee failure. Lumosity.com's memory training thus teaches an end to expansion; signs must not fructify, become spaces or places. The contrast is with Perec's memory exercises, which encourage just such an imaginary expansion. Moreover, in Perec's work, the world of the self is rather consciously opened to others through the act of writing that makes of the intimate memory something that is to some degree public—a mode of publicity very different from the indifferent public rehearsal required by the game.

The endotic, viewed as a "playful" mode of interfacing with the self, takes very different form here[40] from the kinds of memory work demanded by Lumosity.com. Perhaps the connection between the two seems obscure, but Perec's intimate poetics of memory and the kinds of personalized memory games Lumosity.com runs are both versions of what might be termed personal techniques of artificial memory. Moreover, while both rely in part on sequence, they are also heavily dependent on techniques of spatialization, key elements of the art of memory, as it has developed over many centuries. As Frances Yates notes in her authoritative account of the art of memory, the artificial improvement of the natural memory function, at once an art and a technique,[41] was said to have begun as virtuoso performance by the lucky Simonides, a Greek who was able to recall, by "placing" them around the banqueting table at which they died, the victims of an unfortunate act of divine revenge[42] (the insult was a perceived deficit of the required attention being paid . . .). Yates, for whom the interplay between science and magic was crucial, traced the early stories of arts of memory of the classical world through to the giant memory palaces of the Renaissance, where temples, palaces, entire cosmic ordering systems, became intelligent systems for storing and retrieving memories. These were somewhat remote from the daily business of artificial memory (e.g., mnemonics as a virtuoso trade), but one of the startling aspects of Yates's account is the way in which these memory systems required at once an inward journey into the imagination, and the replacing of memory into a far vaster—even cosmic—universe and its systems and rules; an absorption into shared symbolic worlds and highly elaborated worldspaces far beyond the private individual.

Here, then, are three forms of memory work, all of which involved a mode of systematic inward interrogation that is in some way or other automated. The distinction I want to draw concerns memory work as space making or as space constraining, as connecting or withdrawing from shared conceptions of world and place. I also want to underscore the intricate connections between systems of knowledge and understandings of place, and between memory making and place making.

SELF RENOVATION: TIMES AND PLACES

Everyday life: "what lags and falls behind." [43]

For everyday life theorists (among them, Blanchot, who I have cited here), the temporality of the everyday has a certain looseness, it is what "what lags and falls behind," and this may be how it defends itself from various forms of appropriation, from being fully matched with various cultural and social fields and their expectations and value systems, for instance. For Pierre Bourdieu, the structure of the game offers a model for understanding the social world—which is viewed as a series of overlapping fields, games of pre-structured chance, where players with particular embodied skills and abilities (*habitus*) compete against each other.[44] For Bourdieu, however, the model does not equate to the world; structure and agency do not map seamlessly together because the dynamism of the social world in operation (in practice) produces constant struggles, processes of disjuncture, and re-adjustment. This is an analysis based on domination and differential capital, and Bourdieu, exploring the rapidly shifting social structures of the late twentieth century, and the demands they make, writes of the moment when the individual *habitus* cannot keep up with changes in the field—and frames this dynamic as "lag."[45] The looseness of which Blanchot writes is anathema to a world demanding increased performance in all fields.

Of course, this division, between a loose time and a tight one, is already familiar here, at least in relation to memory; lag is that space into which Perec would like to fall, a great significant world, the interior palace that parallels the architectural (if imaginary) memory palace of Yates, that might be roamed within. In such roaming, time might be both found and lost—and also treasures might be found. In contrast, these smartness games in their operation and in their address, reify the present, and the ability to match human processing speed with the speed of the now.

These games speed users up (or, at least, are intended to). They make users operate faster and in synch, perhaps, with what the world (and particularly the information society) demands. Brain training, determinedly intimate and entirely self-centered, a quintessential form of work on the self, like cosmetic surgery and personal genomics (alongside both of which it has risen), is above all a work of renovation. It is an upgrade, a restoration designed to bring the lagging body up to speed with the demands of its environment, to make it more competitive *in the field*.

For the brain trainers, the lag-free life, beating the field, staying up with the requirements of the field, is posited as a desirable goal, a key element of the work on the self that has become the normalized route to advancement and well being in contemporary society. Smartness games offer technology designed to "catch up" users' skills, to update their *habitus*, so that they can cope more effectively with the demands of the contemporary networked world, with the new public space in all its digital, fragmented, multitasking, speeded-up forms. The desire for personal speed up parallels that more narrow desire for lag-free communication, for an end to "temporal disturbances in the flow of communication" (thus, Mia Consalvo defines lag in relation to online activities as a problem caused and resolved by computers),[46] but also translates it into the realm of embodiment and the actions of bodies in space.

Finally, it is useful to note that while lag might be perceived to be a social or at least an interpersonal problem (a question of performance in a shared field of action), the solution offered by brain training is personal; what is being offered is a technological fix—perhaps a cheat, something always intrinsic to the claims made for memory improvement, and cognitive improvement in fact, as well as beloved of gamers with no time.

UNCONVIVIALITY?

These games not only promise to renovate users' *habitus* so that they might perform better in future contests, they also help to turn space—public or private—to users' own immediate ends. Which raises the question of what happens to "the public" and to public space as that which is made a place through interaction with others, when the player plays? There are tensions between the public locations of mobilized practices of personal gaming—and sometimes also tensions between those who inhabit one or both of these worlds. Michael Bull has explored this issue in relation to MP3 use—which might cool down space for those shut out but remaining co-present,[47] and there are parallels to be made between listening "elsewhere" and the kind of play discussed here. Of course, degrees of presence, or the "presence availability" of one individual to another, whether this is gauged in a phenomenological register, or in relation to technological possibilities for various forms of connect or disconnect (the capacity to look away in another space) are not determined by technology alone, but are, as Nicola Green points out (with a direct reference back to Goffman and the mode of civil inattention as a form of courtesy), also conditioned by social categories and cultural formations.[48] Clearly, however, what is at issue is not simply a symbolic re-arrangement of attention, a form of chilling civil inattention maintained only through social codes, but something involving specific technologies, bodily dispositions, and sustained (trained) shifts in *habitus*. This shift in the quality of shared space might contribute to the relatively rarely articulated, but nonetheless often evident, hostility to public gaming. This is part of a hostility or ambivalence, focused around technology, that even if also articulating a broader concern, needs to be taken seriously and explored (if this has not been done within the horizon of game studies it is perhaps because the field in general is, perhaps for good reasons, extraordinarily defensive about the cultural form it explores).[49]

Forms of attention paid to the self through playing these games might simultaneously produce a form of contempt for, or indifference to, the common use of public space. Readdressing the small world of social interaction through the critical optic of everyday life and the production of space as a critical social production, the question of civility and interaction might be mapped onto conceptions of common space—and the demand for the right to make common use of common space—set out by David Harvey as an explicitly political right: as a politics of space and place.[50] This doesn't (either in Harvey's work or here) amount to a demand for nostalgic heterogeneity: for the collapse of the complexity of place, a faux reconstruction of "one space," but it might provoke consideration of questions of what Paul Gilroy termed conviviality[51] and defined as a form of engagement that is an "ordinary feature of social life"[52] which is explicitly considered as an "everyday, ordinary virtue."[53]

Considering race, Gilroy's concept of conviviality is that form of connection or interaction in space whereby "the strangeness of strangers goes out of focus and other dimensions of a basic sameness can be acknowledged and made significant."[54] In her consideration of Gilroy's project, Anker helpfully underscores the moment of quasi Brechtian distanciation that is included in this idea—Gilroy seeks a mode of engagement between people in the social world that allows for a way of looking "with fresh eyes."[55] Here, though, playing games of self-improvement, the eyes are downward, are consciously not-looking out but in. In these kinds of activities, the strangeness of strangers is never *addressed*. Indeed, despite sharing space, strangers are not encountered, except as distractions to be eliminated—or as statistical points on a performance chart. We might say these games do not assist in the organization of potentially convivial relations as relations of place. If they are not hostile to others, then they are markedly unconvivial. In their concentration on the self, not the other, the aim and effect of the games produces, as a principle, not engagement but the mirror of narcissism; the rehearsal studio with its mirrors, rather than the flaneurism of the dérive, with its distanced, but still oddly engaged, mode of connection.

THE CITY OF GLITCHES

In a study based on pervasive mobile use, the urban theorist Anthony Townsend explored the rise of the "telepathic city"[56] asking how networked connections and the flows they organize come to define the city and communicational standards become those by which it might be judged. I have argued that, as a mode of turning away and turning in, this kind of work on the self, at least, may actually outrage, rather than confirm, that "virtually compulsive"[57] demand for intimacy (with others) that Miller describes, that might people this telepathic vision. Or, at the very least, it might reveal a disjuncture between intimate communion with technology (as a mode of changing place) and communication. The confirmation, the necessary audience for the production (or performance) of the self that is said to underpin permanent communication, is in these games found in the mirror of the better-self glimpsed in the games, in the engagement on the self, which, undertaken now, is also for interaction "later." And it is germane to note in this context that even the competition is freeze dried: there are ways of measuring performance against others in these games but only as statistical norms against which to measure yourself (the public as publicity in the old sense of the word).

Mobile gaming might thus work against a normative demand for the kind of "connected presence" that increasingly wants (some part of you) to be "always on"—even as it promises to make you ready to be "just-in-time" or up-to-date. The activity of play, certainly the lone play explored here, creates temporary semi-enclosures, coalescences of virtual and physical locations, brought into being through intense but temporary pulses of activity, which are characterized by their relative isolation and their partial, directionally uneven, temporally complex, separation from the surrounding environment. These are glitches sewn into the larger informational and social fabric.

Thus, it can be argued that what Townsend called the telepathic quality of the city, here re-adjusted to signal the capacity of the city to listen to itself, is reduced in these game zones, is also felt unevenly by players and non-players in their vicinity, and perhaps is felt to be less necessary by those who have retreated, in full public view, into their own rehearsal spaces. Something that these games give those who play them is their own (sense of) space and their own (sense of) time, the illusion perhaps that the world is a playground of their own. This is also that old illusion, proffered by digital technology—and given for real, even as it is also frustrated—of control.

STRATEGIC OPTIMISM?

Finally, I recognize that these are trivial games, trivial pursuits. And I am not against these games; in fact, to a very limited extent, I enjoyed playing them. Nor do I want to subscribe to a pious form of return to an organic "one space," nor to an uncritical mode of anti-computing. This piece has, however, sought to show how these training games, are, by virtue of the relations they prosecute between spaces (the place of the game, and the place of the world-game), and between people (the other who is statisticalized as opponent inside the game, and interpellated as distraction, when found outside of it), seeking to train us, or suggest to us, an ideal use of space, defined as a way of practicing place "efficiently" and in our own best interests.

Another educational game, but one that implies a different relationship between the game, the place, and the life, was played recently in South East London where students protesting against cuts in England have been re-thinking the location of the university. Among the somewhat situationist activities of the University of Strategic Optimism[58] was the mounting of five minute public lectures in the private-public spaces of local banks, which were transformed for a moment, from transactional spaces for private interactions, into common sites for a common learning project.

NOTES

1. In demonstrations in London around Christmas, 2010, students and lecturers making a claim for a particular kind of education, also made a claim for the streets (and for St. Stephen's Green, outside the English Parliament) as a common place. Hence the adopted slogan: "Whose streets? Our streets!"
2. Lumosity.com.

3. Charles Dickens, *Hard Times* (London: Penguin, 1985).

4. John Urry, *Sociology Beyond Societies: Mobilities for the Twenty-First Century* (London: Routledge, 2000).

5. Vincent Miller, "New Media, Networking and Phatic Culture," *Convergence* 14 (2008): 387–400.

6. Miller, "New Media, Networking and Phatic Culture," 338.

7. Caroline Bassett, "Up the Garden Path," *Second Nature* 2 (2009). In that article, I produced seven bulletins from the field of play, an inventory recording the experience of playing these private mind games in a series of different public spaces.

8. On narrativity and the ludic, see, for example, Bernard Perron and Mark J. P. Wolf, eds., *The Video Game Theory Reader* (New York: Routledge, 2009).

9. Georges Perec, *Species of Spaces and Other Pieces*, trans. John Sturrock (London: Penguin, 1997).

10. See, for instance, Perron and Wolf, *Video Game Theory Reader.*

11. For instance, qualities such as rich verisimilitude and rich interaction sustain a series of claims for the aesthetic, popular, cult, commercial, or creative significance of gaming/the game.

12. Stephan Clucas, "Cultural Phenomenology and the Everyday," *Critical Quarterly* 42 (2000): 8.

13. Perec, *Species of Spaces*, 211. But see also Goffman on the importance of the trivial, cited in Candace West, "Goffman in Feminist Perspective," *Sociological Perspectives*, 39 (1996): 365.

14. The original temporary autonomous zones (TAZ) were described as such by Hakim Bey as possibilities in the early days of the web and shortly afterwards declared dead. See Hakim Bey, *The Temporary Autonomous Zone, Ontological Anarchy, Poetic Terrorism* (Brooklyn, NY: Autonomedia, 1985, 1991; also at http://hermetic.com/bey/taz_cont.html)

15. Daren C. Brabham, "Crowdsourcing as a Model for Problem Solving: An Introduction and Cases," *Convergence* 14 (2008): 75.

16. See, Caroline Bassett, "How Many Movements?" in Michael Bull and Les Beck, eds., *The Auditory Culture Reader* (Oxford: Berg, 2003), 343–356.

17. Cracking the rules of a game world and cracking the system that solves a puzzle are parallel processes operating at different system levels. The connection might be explored in relation to protololic layers. See Alex Galloway, "Playing the Code, Allegories of Control in Civilization," *Radical Philosophy* 128 (2004): 33–40.

18. Early net discussions, particularly those focusing on identity and performance (rather than textual analysis), prioritized a division between the virtual and the real. See Elizabeth H. Bassett and Kate O'Riordan, "Ethics of Internet Research: Contesting the Human Subjects Research Model," *Ethics and Information Technology* 4 (2002): 233–247. Later work, in response, tends to downplay "life on the screen." See for instance: Helen Kennedy, "Beyond Anonymity, or Future Directions for Internet Identity Research," *New Media and Society* 8 (2006): 859. However, the question of this activity, and its framing, remains important in exploring the dimensions of contemporary place. Focusing on questions of place and acknowledging the tensions between presence and what might be termed attendance, exposes this.

19. Roger Silverstone, *Television, Technology and Everyday Life* (London: Routledge, 1994).

20. The apparent assimilation of these two traditions, which hides what are conflicts as well as over-laps, is renewed in interesting ways in relation to locative media, where the magic of the new and its prospects, very prevalent in early years of the net, is refreshed so that different ideological readings can

temporarily be supported by the same prospects. The situationist dérive and the neoliberal world of the information economy are peculiar bed-fellows but sleep together nonetheless. An echo of this disjuncture is identified in relation to the work on the self by Louis McNay. The latter points to tensions in Foucault's later writing between the project of the self as enterprise and the ethics of the self as practical resistance to domination. See Louis McNay, "Self as Enterprise: Dilemmas of Control and Resistance in Foucault's *The Birth of Biopolitics*," *Theory Culture Society* 26 (2009): 55.

21. Erving Goffman, *The Presentation of Self in Everyday Life* (New York, 1959), 244.
22. Candace West, "Goffman in Feminist Perspective," 353–369.
23. Candace West cites Nancy Henley and also Dorothy Smith, as exceptions to this trend. See Candace West, "Goffman in Feminist Perspective," 353–369. West also argues that Goffman understood the importance of the relative degrees of symmetry or asymmetry operating between actors and between various stages/sites on which action may be staged. It might be added that he did so in relation to a politics of place and space because it was upon various stages that these performances of the self were undertaken.
24. Mark Selzer, among others, notes that Goffman was clear that civil inattention is a form of courtesy. See, Erving Goffman, *Relations in Public* (London: Penguin, 1972), 385, cited in Selzer's "Parlor Games: The Apriorization of the Media," *Critical Inquiry* 36 (2009): 100–134.
25. Selzer, *Parlor Games*.
26. Jan Huizinga, *Homo Ludens* (London: Maurice Temple Smith, 1970), 26–30, cited in Roger Silverstone, *Why Study the Media* (London: Routledge, 1999), 61. See also: Hector Rodriguez, "The Playful and the Serious: An Approximation to Huizinga's Homo Ludens," *Games Studies* 6 (2006), http://gamestudies.org/0601/ articles/rodriges.
27. Selzer, "Parlor Games," 101.
28. Margaret Tan Ai Hua, "The Paradoxical and the Reversible: Towards a Critique of the Intelligent Nation 2015 (iN2015) Masterplan and Pervasive Computing in Singapore," unpublished Doctoral Dissertation, National University of Singapore, 2011.
29. Selzer, "Parlor Games," 101.
30. The distinction is Deleuze's. Gilles Deleuze, "Postscript on the Societies of Control," *October* 59 (1992): 3–7.
31. See Jesper Juul, "Games Telling Stories?—A Brief Note on Games and Narratives," *Game Studies* 1 (2001) http://www.gamestudies.org/0101/juul-gts/
32. See, for example, Nicola Green, "On the Move: Technology, Mobility, and the Mediation of Social Time and Space," in *The New Media Theory Reader*, eds. Robert Hassan and Julian Thomas (Maidenhead: Open University Press, 2006), 249–266.
33. The contrast between this invitation to productivity and media activities offering ways to use these spots of time idly—perhaps to skim what Joke Hermes called "put downable" media—is marked.
34. Marc Augé, *Non-places: Introduction to an Anthropology of Supermodernity* (London: Verso, 1995).
35. Gemma San Cornelio and Elisenda Ardèvol, "Subjective Cartographies and Augmented Spaces; Practices of Place-making through Locative Media Artworks," *Communication* (2011, forthcoming).
36. As Mattelart and Mattelart put it, the clear distinction between this and the French tradition is that the latter has "emphasized" the "mechanisms of socialization of communicating devices," while French critics of social

geography, such as De Certeau, also clear that actions are still possible, are also "driven by the intimate conviction that the mechanisms of subjection are ever present" (Armand Mattelart and Michelle Mattelart, *Theories of Communication* [London: Sage, 1998], 126–127). One key difference between the French and US approaches concerns the degrees of self-reflexivity and agency accorded to social actors or social subjects within the structural constraints of an over-arching system.

37. The everyday is "where the marvelous exists." Ben Highmore, cited in Seth Giddings, "A Pataphysics Engine, Technology, Play, and Realities," *Games and Culture* 2 (2007): 392–404.

38. Selzer thus follows his consideration of parlor games with an exploration of anteriorization in relation to Ishiguro's *Never Let Me Go*, considered as a murder mystery.

39. Frances Yates, *The Art of Memory* (Chicago: Chicago University Press, 1966), 6–7. The arts of memory have at various times been deployed in forms of virtuoso performance, but have also always been offered as a skill to develop in private. In more recent times an improved memory, acquired through memory work, has been meant to look natural, like a talent. Thus, memory improvement products of the late twentieth century flattered innate ability ("IQ of 200 and can't remember") whilst also promising discretion.

40. For Perec, this narrative expansion is accessed through various forms of systemized rules; playful modes of data interrogation. For discussion, see Caroline Bassett, "How Many Movements?" *Open, Cahier on Art and the Public Domain* 9 (2005), 38–48; also at http://www.skor.nl/article-2854-en.html.

41. See Heidegger on techne and poesis in the "Question Concerning Technology," in Martin Heigger, *Basic Writings*, ed. David Farrell Krell (London: Routledge, 1993).

42. Yates, *The Arts of Memory*.

43. Blanchot, a French thinker in the same tradition as Lefebvre, described the contradictory nature of the everyday as "platitude (what lags and falls behind, the residual life with which we fill our trash cans and cemeteries: scrap and refuse)," whilst also arguing that "this banality is also what is most important if it brings us back to existence in its very spontaneity and as it is lived" (cited in Stephen Clucas, "Cultural Phenomenology and the Everyday," *Critical Quarterly*, 42 [2000]: 9–10).

44. Pierre Bourdieu, "Some Properties of Fields," *Sociology in Question*, trans. Richard Nice (London: Sage, 1993), 72–77.

45. Bourdieu, "Some Properties of Fields." Bourdieu's world might be considered closed (structure and agency informing each other), but this is to neglect what he insisted was the opening, the element of dynamism, and new ones are continually consecrated. It might be said that it is precisely because fields are not models of each other, are not fractally connected, that this mismatch, this opening, exists. In his model, not only do bodies lag and fail to conform, fields overlap, rise and fall.

46. Mia Consalvo, "Lag: Language and Lingo: Theorizing Noise in On-line Spaces," in Bernard Perron and Mark J. P. Wolf, *The Video Game Theory Reader 2* (London: Routledge, 2009), 303.

47. Michael Bull, "No Dead Air! The iPod and the Culture of Mobile Listening," *Journal of Leisure Studies* 24 (2005): 343–355.

48. Nicola Green, "On the Move: Technology, Mobility, and the Mediation of Social Time and Space," in Robert Hassan and Julian Thomas, eds., *The New Media Theory Reader* (Maidenhead: Open University Press, 2006), 249–266.

49. See, for instance, Espen Aarseth's assessment of the field of computer game studies in "Computer Game Studies, Year One," *Game Studies* 1 (2001): 1–4.

50. "We need to find ways to build a dialectics of politics that moves freely from the micro to the macro and back again" (David Harvey, *Spaces of Hope* [Edinburgh: Edinburgh University Press, 2000], 52).

51. Paul Gilroy, *After Empire: Melancholia or Convivial Culture* (London: Verso, 2004).

52. Elisabeth Anker, "Review Essay: *National Love in Violent Times*," *Political Theory* 36 (2008): 765.

53. Anker, "Review Essay," 135.

54. Paul Gilroy, *Postcolonial Melancholia* (New York: Columbia University Press, 2006), 3–4, cited in Anker, "Review Essay," 765.

55. Gilroy, *Postcolonial Melancholia*, 135, cited in Anker, "Review Essay," 765.

56. Anthony, Townsend, "Connecting to the Future Seoul Searching—Cybernomads and the Ubiquitous City," *Receiver* 13 (2007): http://www.vodafone.com/flash/receiver/13/articles/pdf/13_02.pdf.

57. Anthony Giddens, *The Transformation of Intimacy: Sexuality, Love and Eroticism in Modern Societies* (Cambridge: Polity, 1992), 97.

58. University for Strategic Optimism: http://universityforstrategicoptimism.wordpress.com/

Part III

Urbanity, Rurality, and the Scene of Mobiles

7 (Putting) Mobile Technologies in Their Place

A Geographical Perspective

*Chris Gibson, Susan Luckman,
and Chris Brennan-Horley*

INTRODUCTION

This chapter critically engages with the proposition that mobile technologies challenge place as a "proper, stable and distinct location,"[1] drawing on the rich history of theorizing place in geography. Our engagement with this proposition draws critical attention to the assumption that place was ever "dominantly" understood as stable or distinct. Indeed, geographers have argued for decades that conceptions of place must move beyond ontologies of strongly bounded geographical scales.[2] While the assumption of "place" as a neatly bounded category might linger in some cultural history and cultural studies research (especially that having recently "discovered" the spatial), as Kevin Dunn argues, "cultural geography's engagement is both older and deeper"[3]—producing enduring disciplinary legacies and anxieties, as well as key insights. Among the legacies is a suspicion of super-organic conceptions of place that assume inert conceptions of landscape—place as "blank sheets onto which culture was written." Much of this kind of thinking, we admit, does infuse contemporary accounts of the transformational effects of technological advancement on experiences of place—place as unmoving, a rudimentary container until it is transformed through exciting socio-technological change, or what Goggin calls "the technological sublime."[4] As cultural geographers we are uncomfortable with this, retaining "an anxiety not to replicate the problems of environmental determinism, or to reduce space to a container"[5] in the manner that characterized cultural geography in the 1920s through to the 1940s.

Against the backdrop of a prolonged disciplinary obsession with "place," geographers of different persuasions have since the 1970s developed more reflexive and dialogic theories of place—that emphasize the dynamic and contingent *production* of location—with place having no privileged ontology and instead being viewed as the locus of intersections of geometries of power.[6] For humanist geographers in the 1970s,[7] place was "concerned with individuals' attachments to particular places and the symbolic or metonymic quality of popular concepts of place which link events, attitudes, and places to create a fused whole."[8] Place is seen as a dynamic psycho-social construct.

Around the same time, geographers of Marxist political economic persuasion emphasized technologies of time-space compression[9] that emerge out of the capitalist imperatives to disperse and reinvest surplus capital—hence capitalists re-scale spatial relations through globalization and strategic localization, in effect harnessing places as the "engines of capitalism."[10] For Marxist geographers, place is where capitalism organizes and coalesces facilities, infrastructures and investment—literally the places of the transformation of nature, of production and accumulation. For geographers, then, places are *produced* rather than merely extant.

Even more recently, fresh geographical perspectives have emerged, informed by poststructuralism and theories of embodiment, performativity and affect, that position people, technologies and material qualities of places near and far in assemblages, circuits and networks—complex *spatializations* with less clear arguments about causality than Marxist geographers, but sharing an ontology that recognizes the dynamic, even unstable, character of place, within which everyday "senses of place" are constituted.[11] Such perspectives imagine human and non-human actors, ideas and materials awhirl in decentered webs of intimate and less-than-intimate relationships—opening up the possibility of simultaneously conceptualizing physical geographical matter (biospheres, plants, animals, fluvial systems, water and nutrient cycles, coastlines, soil, weather) and human technologies and activities—rather than persist with the modernist separation of human and non-human entities into the discrete sciences of botany, geomorphology, anthropology, economics, and so forth.[12] From such perspectives, it is the jostling and interactions of a multitude of "things" that make "place" what it is. Rather than assume that it ever was "stable" or "proper," place comes to be understood as "becoming,"[13] a continually unfurling phenomenological mix.[14]

Where is the place for technologies—such as locative media—in all of this? All manner of human technologies (from railways to email) warp both our material experiences and discourses of space and time, just as those technologies emerge and operate within assemblages of human power, cultural proclivity, and commercial imperative. For example, in Bill Cronon's classic work on Chicago, the railways (and to a lesser extent, the telegraph) played a critical role in newly regulating time, transgressing seasonal limitations on travel and production, and commensurately transforming the ecological space of the American Prairies in the 1850s. Precise time zones and absolute measurement of standard times were necessary to coordinate train timetables to and from Chicago (and to standardize records of telegraph transmissions), surpassing intricate, place-specific and longitudinally varying measurements of time based on overhead positioning of the sun. Further, Prairie farming systems were extended and intensified by railways: "no earlier invention had so fundamentally altered people's expectations of how long it took to travel between two distant points [. . .] railroad and telegraph systems would expand in tandem, often following the same routes, and together they shrank the whole perceptual universe in North America."[15] In the same way, mobile media emerge from, and dialectically transform, places in complex ways—in some places more than others: "place does indeed persist in and through networked mobility,"[16] but exactly how remains moot.

In this chapter, we embrace this dynamic and intersecting geographical perspective—to situate mobile technologies *in place*, as it were. We seek to move on from Morse's proposition that media technologies challenge place as a "proper, stable and distinct location,"[17] arguing that theorization about how mobile technologies might transform places itself needs to put considerations of geography at the center of things—rather than at the end of a simplistic chain of causality that link people to technology, to (reconfigured) place. Our argument rests on two parallel observations.

First, physical geography is an *active* agent in the "new" landscapes made possible by mobiles—just as it was in earlier attempts to build railways or telegraphs across continents. In terms of physical infrastructure, the experience of Internet and phone service varies greatly across space. Mobile technologies from the most basic, such as phone calls and short message service (SMS), through to Internet-enabled smartphones are entirely dependent on network coverage across topography. Innovative smartphone developments in the areas of global positioning system (GPS) navigation and augmented reality can be rendered useless depending on one's location, terms of service, and handset capability. But also, certain kinds of cultural uses of mobile technologies rest on underlying physical geographical qualities, such as proximity, in actively shaping everyday experiences of place.[18] Locality is itself mobile, or as Larissa Hjorth has argued, it is "in the post;"[19] but place is also embedded in the very infrastructures that deliver that post. Physical landscape is not separate from humans, or inert—but is instead one among many powerful actors, shaping the possibilities and limitations of human technological innovations.

Bringing attention to the physical qualities of place may seem a basic point to make, but even in commentaries in media studies well-attuned to the complexities of geography,[20] non-human dimensions of place often only figure as a stage on which human practices (such as domesticity, community, identity) are built or contested. Just as it was in the days of the telegraph and railways, today, too, the agency of landscape as a key non-human actor in the roll-out of mobile technologies is glaringly apparent in the difficulties encountered in delivering services across the large, often sparsely populated landmasses of countries such as Canada, Australia, China, and Russia.

In our case study of Darwin, Australia, which will be discussed at length, the physicalities of geography (i.e., weather, distance, topography) fundamentally insert themselves into the technological mediation of place by humans. Further, and linked to this, the cost of access in rural and remote areas can be highly prohibitive for all but commercial use. In a similar manner to the uptake of Internet in Australia, uneven geographical provision of telecommunications can exclude certain socioeconomic groups from access to emerging technologies.[21] Wireless connectivity is far from ubiquitous.[22] These are not mere superstructural issues to be left to infrastructure engineers and service providers—although the ideals of universal service provision and digital commons remain worthwhile ongoing sites of struggle[23]—but rather constitute important features in an actor-network that links humans and mobiles across space. Cultural analysis of mobile use in place therefore needs to engage with how people navigate physical geography, uneven access, capacity to pay, and the creative limitations of current, private-sector service delivery platforms.[24]

Second, as Goggin intimates,[25] despite the lauded potential for mobile phones to transcend space, in people's quotidian realities how significant is this impact as determined by place? Those advancing notions of the "technological sublime" (usually telecommunications companies and government agencies, but also to some extent academics, too) tend toward two polar arguments: first, the erasure of physical geographical boundaries and barriers by new locative media (the "end of geography" thesis); and second, the enlivening of geography by geocoding everyday life through GPS chips and tagging functions (the "triumph of geography" thesis).[26] But both rest on a problematic assumption of humans liberated from—or in masterful control of—place. Place is either superseded or celebrated, but in both cases is conceptually a backdrop to the wonders of human ingenuity. No more is this visible than in the ubiquitous reliance of smartphones, GPS apps and other mobile devices on standard platform mapping websites (such as Google Maps) and base layers. These platforms use maps based on western, scientific cartographic practices—maps as the ubiquitous, seemingly transcendent "view from nowhere,"[27] belying their power, selective construction and Panopticon qualities.[28] Location is increasingly pervasive in mobile technology—but the version of locality appropriated is flat. Physical geography is but a base layer on top of which all manner of human user-generated content can be loaded. As Sanders argues:

> What is troubling about this is the extent to which we really believe that we live in a placeless, totally globalized world, characterized by the ability to move through time and space faster and more. This idea has gone virtually uncritiqued. The irony is that at the same time [. . .] more than ever we are now enslaved to geography—the geography of the Cartesian map. The new communication technologies, often referred to as locative media, claim to overlook the conventions of the base map and "imagine an alternative organisation" of space but in reality they actually reinforce the most traditional conventions regarding location.[29]

In our recent study of the everyday experiences of Darwin's creative industry workers (an Australian Research Council Linkage Project principally concerned with city cultural planning issues, but that considered technological mediation of creative work as a matter of course), it was noted how infrequently mobile technologies were referred to as enabling creative professionals to go about their work. Rather, Internet accessibility was as much of an issue for creative workers in Darwin as mobile reception. Given the recent convergence between Internet and mobile services (mobile broadband) their opinions toward the Internet, as well as mobile phones, are illustrative of wider discourses regarding technology and service delivery in remote settings. Located as it is in Australia's north, closer to key Asian cities than the urban centers of Australia, Darwin is a location defined by a sense of "remoteness;" a feeling differently enabled but not transcended by the emergence of new transport and communications technologies. Darwin is a high-tech place, as befits a northern capital with strong mining and military industries. Indeed, as a direct result of its unique location it has been on the cusp of telecommunications innovation and linked to the world ever since the overland telegraph went through Darwin on its way back to the "mother country." Prior to this, pre-colonization, the indigenous people of the area were in regular contact with Asia via regional trade routes. All these histories see geography as an active agent in technological negotiation. "Remoteness" is a substantial issue for creative practitioners who have a profoundly strong sense of place in terms of what location means to them and how it impacts on their lives. And yet, despite on the whole being mobile phone and Internet-savvy citizens, for creative industry workers information technologies (including mobiles) were taken for granted items—not life changers/savers. This is not to say that people lack sophisticated ways of negotiating contact across distance, but rather that capabilities to enroll mobile technologies in place-transforming ways are themselves shaped by geographical circumstances. In the remainder of this chapter we draw on this particular project to illustrate our point.

DARWIN, AUSTRALIA:
A TELECOMMUNICATIONS FRONTIER?

Although a capital city, as already suggested, Darwin is clearly defined—in the national psyche at least—by its remoteness from the vast bulk of the Australian population. That remoteness may be difficult for international readers—and indeed many southern Australian readers—to grasp. It is 1,500 kilometers by road to the nearest substantial town (Alice Springs, which has a population of only 25,000) and 3,000 kilometers to the nearest state capital city (Adelaide). London is closer to Cairo in Egypt than Darwin is to Sydney, Melbourne, or Perth. As such, telecommunications technology loomed large in our research as a feature of going about one's business in Darwin; access to IT is a core need of Darwin's creative sector, as well as the larger mining, defense industries underpinning the region's economy. As we shall see, it is another enabler and inhibitor, used to overcome the "tyranny of distance"—particularly for sourcing material not readily available through local avenues—but also, depending on one's resources and places of collaboration, it underscores geography's place as a non-human actor underpinning take-up and use of mobile technologies. Importantly, too, at an international level, the Internet (rather than mobile phones *per se*) helps give expression to Darwin's place within international networks (i.e. given its proximity with Asia and the amount of business done there), above and beyond any mitigation of distance from the Australian capitals.[30] The importance of links to close international networks, as well as the many smaller, even more isolated communities of Australia's north (including Aboriginal remote townships), was evident from a mapping exercise carried out with research participants.

Here links were repeatedly made between Darwin and remote communities across the Northern Territory. Business with each of these sites habitually necessitates expensive mobile telephony: international calls and satellite phones. Given that the Internet, however accessed, enables more affordable communication within the realities of this geography, it is hardly surprising that issues of broadband speed and reliability, plus access, remain key concerns. This situation only becomes all the more pertinent as smartphones

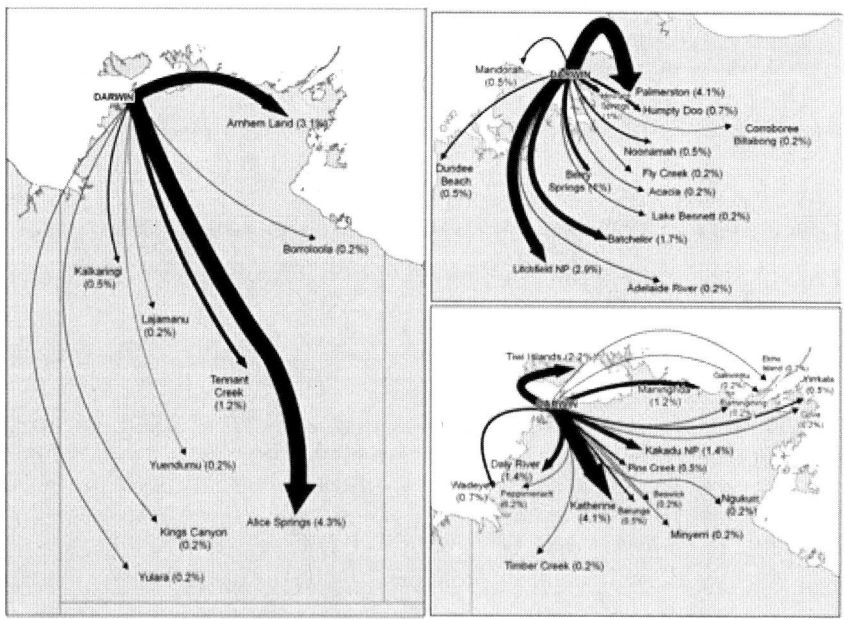

Figure 7.1   Creative practitioners' connections from Darwin to other Northern Territory locations, expressed as a percentage of all work linkages within and beyond Darwin, 2008.

become the norm and the quality of access to the Internet increasingly determines the ultimate usefulness of one's mobile phone handset.

ACCESS, SPEED, RELIABILITY: PLACING MOBILE TECHNOLOGIES IN DARWIN

Despite governmental commitments, issues of broadband speed and reliability, mobile phone reception, plus access—especially with collaborating partners in rural, isolated and poor and poorly serviced parts of the region—remain key concerns. Notwithstanding technological advances and the widespread use of satellite phones in the Northern Territory, geography remains a key non-human actor in Darwin's communications landscape. As described by one respondent, whose job is to coordinate and market arts projects:

Q: What are some of the things that make working in a place like Darwin challenging but also rewarding?

A: Challenging: basic communications, like getting information out to regional areas that don't have Internet, or, you know, the fax might be down, or there might be some kind of sorry business going on, so no one's talking to anyone for a couple of months. Geographical distance, trying to have meetings with people that are four hours away and trying to get them to come here or me go there. So, yeah, the actual space, the geographic space and technology is hard also.[31]

Even within the city of Darwin itself reception is patchy and geography thus a powerful actor, as illustrated in a multimedia and marketing professional's experience:

Q: What impediments do you perceive there to be here in relation to inhibiting creativity in Darwin?

A: Broadband speed. It's shocking . . . compared to every other capital city in Australia, we have the worst broadband speed . . . The people out here in the suburbs and that, or even in the city, it's just not fast enough and it's not good enough for everyone . . . I bought a house, 200 meters just down that way, and when we moved in, we found that we couldn't get Internet access, and as you can imagine in my business, I definitely need Internet access. So I hassled Telstra, you can keep this on record, I don't mind, for about 8 weeks, only 8 weeks later did I get Internet access.

Telstra's own mobile network coverage maps for the region[32] reveal the partiality of mobile reception across the region, concentrated in the Darwin city area, along parts of the one major highway out of town (the Stuart), and in coastal zones (typically those areas important for fishing, military uses, and border protection patrols).[33]

Rarely does coverage extend inland, despite the region being home to many thousands of Aboriginal people living in scattered settlements and communities (and which were important networking sites for the creative industry workers we interviewed, especially those working in the internationally famous Aboriginal arts field—see Figure 8.1 and note 33). Telstra's publicly available coverage maps are created using geographical information systems technologies (GIS), with data on phone tower type and location, but they overestimate the extent of coverage by a considerable amount because of lack of integration of that data with other map layers including topography, vegetation, and built environment, that would more accurately enable maps to reflect on-the-ground reception.[34] The main provider of mobile coverage to the north, the formerly nationalized Telstra Corporation admits as much itself on its website where the organization acknowledges the software's inability to appropriately account for the on-the-ground impact of "certain physical structures or geographic features [. . . which] may block or inhibit coverage [. . . and which] could include [. . .] hills and mountains or even trees."[35] The website emphasizes this is a problem common to all wireless systems.

At the coalface of user experience in the region, such unreliable coverage can reasonably lead to less emphasis on wireless technologies as an essential tool of everyday life. Less-than-perfect issues of service availability were perceived as exacerbating, in the opinion of a graphic designer, a tendency toward Luddite-ism characteristic of the remote frontier setting:

> In terms of digital creativity, it's still . . . a lot of people are still outsourcing to Sydney, still going elsewhere for that creative medium, and they don't have to anymore. And I think that there is a lot more room, especially for re-education of the people that live here. I mean, like I said, there's never ever been anything like us, because we're web based. Lots of businesses here still say that "I don't believe in the web," so there's still that sense of mentality that we're all up against.

But at the same time, reminders of the reality of remote, frontier life were constant. One of our own incidental experiences illustrates this: on one research trip for this project, the major hotel at which we were staying was struck by lightning, knocking out the hotel's entire Internet service. The technician needed had to be flown in from interstate to fix it—a minimum 3 hour flight, which meant that all Internet services were out for over 24 hours. This may have been a minor inconvenience for us as visiting researchers, but for those living with such limitations in an ongoing manner the frontier experience is still vivid, curtailing any sense of a technological sublime.

DIGITAL RESIGNATION AND "MAKING DO:" HUMANS
AND MOBILE TECHNOLOGIES IN THEIR PLACE

Given this situation, why has there been no "telecommunications uprising?" If we examine this question via a discourse of "lack" it could be argued that it is hard to miss that which one has never had, but this would ignore the reality that technology users in places such as Darwin are frequently highly mobile themselves, physically, and are acutely aware of the kinds of connectivity their iPhones afford them in Singapore, Sydney or even downtown Darwin. Perhaps a more useful lens through which to explore this is to consider, following Bourdieu,[36] what digital capital actually looks like in different places, and thus how place is a contingent actor in this relationship. Further, what then is "universal service" to rural and remote communities, and to what degree is this about "making do" literacies[37] and not just about the technology itself? In some instances in our study, the much romanticized "frontier ingenuity" was invoked as an enlivening mode of interaction with otherwise patchy mobile phone and Internet availability—people in remote areas being creative in figuring out ways to "make do." Hence, for Michael Denigan, Darwin personality and seller of Mick's Whips (locally made leather whips):

> Telephone is essential. If it wasn't for the telephone, like we've been living on our block for 16 years and if it wasn't for that you couldn't survive. We used to have a bag phone that we used to attach to the car. When Internet first came out in '96 we got a mate in Darwin to get online for us and then he would relay us the messages and the credit card transactions onto our analogue bag phone which was attached to the car, and then we'd do all the paperwork, make the products and once a week go into town and sell it. We used to go around Australia and do the same thing, visiting remote areas. As soon as you'd come out of the Daintree and be at Cairns your phone would go off with all our orders and all the whips that you'd made in the Daintree or wherever you just pack up and send them off, put the money in your account, see you later. So if it wasn't for mobile phones they would never have been achievable you know.

As Kate Bowles reminds us,[38] rural, regional and remote people are sophisticated knowers of what they don't have, and we need to think about the ways in which rural, regional and isolated families handle the concept of cost as not just a financial picture, but also one where lifestyle and opportunity need to be factored in.

> Another reason for the absence of protest is a measure of acceptance by Darwin residents of geography and what it means in terms of access to, and use of, Internet and mobile technologies, especially in the tropical

wet season. For some, disappointment with mobile media and the Internet was coupled with a particular kind of laconic resignation—a sense of defeat at the hands of non-human actors beyond anyone's control, so the impact of weather on telecommunications features strongly in response to the question: "How does the wet season impact upon your creative work?"

A: "The storms crash our computers. That's a major issue." (Events and Conference Manager)

And:

A: [it only majorly impacts if] we get struck down by a cyclone or something like that . . . everyone that's working here is a Darwin person . . . so we're kind of used to the wet season. Air-conditioning dying can reduce productivity substantially, it increases the whinging, but, the only time [it really impacts] is the storm surges, like having the power outages and things like that, because we just do. As soon as there's a lightning storm around, we don't risk our technology, we unplug, we just take it straight out. But that time is a good chance to catch up, with a glass of wine, have a chat, depending on the time of day, talk our projects through.

Contra the technological sublime, for most of our interview participants, the Internet (more so than for mobile telephony, which was the cause of frustration because of reception limitations) was carefully and selectively adopted within existing practice—on the whole quite fine for allowing people to do what they need to do. Telecommunications policy often assumes the high-end user is everywhere, rather than thinking of this as a niche market in most areas. Mobile and Internet technologies are partially and unevenly adopted for many different reasons. Thus, for a visual artist:

I constantly phone Melbourne for supplies and stuff, and then we have got suppliers here that are kind of like the middle people, and they can help sort that through. The Internet—apparently there are good ways to order online, but I haven't done that yet . . . sometimes I will just fly to Melbourne and buy up, put it on the plane, you know, as long as it's not flammable, and just bring it back.

This is perhaps especially shaped by the cognizance of Darwin creative workers that they are actually quite well-connected compared with their regional partners—both in Australia (e.g., compared to remote indigenous communities) and in the wider region (Indonesia, the Philippines, Timor)—where there is no point investing in advanced mobile technologies or sending large files when dealing with collaborators who can't receive or download them.

In other instances, the exigencies of commercial career-building necessitate the persistence of face-to-face interactions—interactions that simply can't be simulated or vicariously experienced. For the manager of Skinnyfish Music (a music label distributing mostly Indigenous artists, including the enormously successful Geoffrey Gurumul Yunupingu):

> Internet access and, really, all of that type of technology, it has made it much easier for musicians here to actually gain a profile. But without all the infrastructure the musicians here would never have, wouldn't have left anyway. So it's not a question of technology helps musicians stay here, they would have stayed here without it, but with no exposure. Now they stay here and get exposure. We put all our releases on digital download, but, again, the problem is that we need the artist to perform [as] that generates the interest to go and look at what they do. So, again, the issue in terms of world sales is that we need to get an artist to the marketplace to perform. And that's why, you know, with Geoffrey and Shellie [Morris], it actually needs to be something that occurs frequently. They need to be able to go to Europe twice a year for the next five years, and build up a really strong following.

Thus, even with access to technology, distance in Darwin means that place challenges stable and universal conceptions of mobile connectivity.

CONNECTING DARWIN: OVERCOMING THE "TYRANNY OF DISTANCE?"

Underscoring a more conventional vision of mobile connectivity as transcending place, interviewed arts and creative business managers spoke of the key roles of fast broadband and mobile telephony in ensuring their business remained in touch with leading developments from larger interstate firms, enabling contacts to be maintained for professional development opportunities and the like: "access to internet connections, that's probably what keeps me involved in the national understanding of what's happening in the arts." (Art Retailer). More prosaically, for a *NT News* journalist:

> I mean the access to equipment and access to resources has changed entirely with the internet, so I think location is less important these days than it was 10 years ago . . . everyone's sort of interconnected electronically, so, whether it be ordering equipment or getting information, it's completely different, now it doesn't really matter where you live, in Alice Springs or Katherine or Darwin, you've still got access.

Some had witnessed Darwin, the city, change in tandem with technological developments, in ways that suggest a broader, dynamic attitudinal shift in this remote setting. For a Darwin-based architect:

> [the] Internet was hardly used when I left, and now it's one of the dominant modes of communication. Mobile phones were hardly available, so, you know, now you can play games and download movies on your mobile phone, so there's been technologically very large changes, and so you could expect that this place hasn't remained as remote from those influences. But you know, sometimes after being up here for a while, going down [south] and finding these things was a bit of a surprise, it is possible to kind of become a bit of an island up here, but it's less possible now. That's part of the changing economic and social environment.

New communications technologies were thus seen as part of a wider transformation of cultural uses of technology able to challenge traditional vision of isolation, producing a national, and homogenously *urban*, milieu for media:

> I think that has as much to do with Darwin and the Territory growing as an economy. As the population grows . . . I guess we're becoming a lot more like the rest of Australia, in terms of our services and what we have here. But also, you can't deny the digital revolution, in terms of communication and desktop facilities. The way what I do is so much different to what it was, even 10 years ago; it's allowed a lot more sort of merging of media and skills and people crossing over into different areas. So yeah, that's changed, and that kind of coincided with me stepping outside of the mainstream, you know, traditional media. (Independent media producer)

The "end of geography" thesis reverberated—for some—such as the manager of Throb nightclub and *ResideNT* magazine: "because we, everyone, has internet cable or whatever, so we get to see what's happening around the world straight away."

Thus, the Internet, with its cost-effective affordances of international as well as national communication, helps give expression to Darwin's place within communication networks far more so than mobile telephony. This is especially significant given its proximity with Asia and the amount of business done there, and has applicability to other sites where geography facilitates strong international links while the regulatory and commercial infrastructure around telecommunication focuses on the national. International and not just contiguous land-based understandings of geography are key here as these affordances are above and beyond any mitigation of distance from the Australian capitals:

The Internet is certainly a great source of community for many people and it's also another way of positioning your practice or your portfolio of practices within a broader market-place, and I think trying not to always be looking back into the mainland of Australia but to be actually looking out. (Academic and visual artist)

This particular finding does not come from a vacuum. Released in 1995, the *Committee for Darwin Report* (chaired by Neville Wran and commissioned by the Keating government), clearly flagged a future where Darwin's history and geography, coupled with technology and diversified markets, would see Darwin cement its place as Australia's "gateway to Asia." To quote the report:

> [The increasing role of cities and sub-regions in the "intensifying competition" for trade and investment in and with Asia] is one of the most important global trends of the 1990s, and it will be reinforced by continuing changes in communications and transportation technology well into the next century. [. . .] The Committee's approach acknowledges the vital role that smaller, sub-regional units that are more responsive to local market needs will play if Australia is to compete effectively for trade and investment in East Asian markets.[39]

But, like much Australian telecommunications policy in the years since 1994's *Creative Nation* document,[40] the full realization of this potential via decent IT infrastructure has not come to pass. However, this is not to say that the inevitable enhanced linking of Darwin to its near neighbor communities via the low-cost-to-user option of the Internet (as distinct from phones, flights and couriers) has not proceeded at all. Indeed, given the classic "digital divide" which finds many of these near neighbor communities themselves struggling with IT infrastructure, in some ways things are not as bad as they seem. The Internet—much more so than mobile telephony, with its country codes, expensive international calling tariffs and need to confront language translation issues in the immediate moments of phone conversations—has increasingly become the easiest, everyday means through which to nurture nascent pan-Southeast Asian connections. The problem of limited connectivity is much more acute when one is accustomed to, or is dealing with, people in better served centers such as inner-city Sydney, Canberra, Singapore, or Seoul. Patently, in such a context, material geography is far from inert.

CONCLUSION

Rather than suggesting that understandings of place need changing to accommodate the particular experiences of mobile use, we argue that mobile technologies in rural and remote areas remain completely circumscribed by geography, but a contingent sense of geography—one neither distinct nor stable. Furthermore, geography—like technology—also operates within economic and political networks, something Darwinites are only too aware of:

> So, I think that there's increasingly an approach from the federal government to further marginalize remote areas, because they are unprofitable and they're subsidized by the rest of Australia; so your average Sydneysider wouldn't be that impressed about having to pay more for their phone calls, because, there's people out the back of [nowhere] that require phone access [cheaper than the present] phenomenal rates or satellite costs. Australia is the size of Europe, but it can go from a few thousand people per square kilometer to one person every two square kilometers. So I think that western macroeconomic policies can't necessarily be adapted to such a defused economic landscape quite as readily. I think that people need to not quite be so focused on the centers and see how things move throughout it. (Sound Artist)

Technology sits in place, and shapes it, but instead of emphasizing technological determinism over place, we prefer geographically entangled ideas regarding the immediate spatial possibilities of technology. Despite access issues which can easily be perceived as "absences" to those accustomed to dense urban connectivity, underlying what we found was, by contrast, a very rich and tactile geography of the city, experienced by people in their working lives, networks, and experiences of place. Computers were turned off during storms in favor of down-time while the Internet-enabled Skyped with international collaborators; sometimes supplies were ordered over phones, at other times people flew personally interstate to overcome the continued tyrannies of material geographical distance. All these actions reflect a creative negotiation of the physical realities of geography in and through mobile technologies—as well as "making do" with the constraints of those technologies themselves. Thus, in the Darwin area, technology becomes but one more constitute part of the complex and mutually dependent assemblages of geography, climate, history, and peoples, which together structure contingent understandings of place.

NOTES

1. Margaret Morse, "An Ontology of Everyday Distraction: The Freeway, the Mall, and Television," in *Logics of Television: Essays in Cultural Criticism,* ed. Patricia Mellencamp (Bloomington, IN: Indiana University Press; London: BFI, 1990), 195.
2. Yi-Fu Tuan, "Place: An Experiential Perspective," *The Geographical Review* 65 (1975): 151–165; Yi-Fu Tuan, *Topophilia: A Study of Environmental*

Perception, Attitudes, and Values (New York: Columbia University Press, 1990); Richard Howitt, "'A World in a Grain of Sand:' Towards a Reconceptualisation of Geographical Scale," *Australian Geographer* 24 (1993): 33–44; Tim Creswell, *In Place/Out of Place: Geography, Ideology and Transgression* (Minneapolis, MN: University of Minnesota Press, 1996); Sallie A. Marston, "The Social Construction of Scale," *Progress in Human Geography* 24 (2000): 219–42; Doreen Massey, *For Space* (London: Sage, 2005).

3. Kevin Dunn, "A Comparative Genealogy on Place: Cultural Geography and Cultural Studies," *Geographical Research* 44 (2006): 423.

4. Gerard Goggin, "SMS Riot: Transmitting Race on a Sydney Beach," *M/C: A Journal of Media and Culture* 9 (2006), http://journal.media-culture.org.au/0603/02-goggin.php.

5. Dunn, "A Comparative Genealogy," 425.

6. Doreen Massey, "Geographies of Responsibility," *Geografiska Annaler* 86 (2004): 5–18.

7. For example, see Yi-Fu Tuan, *Space and Place: The Perspective of Experience* (Minneapolis, MN: University of Minnesota Press, 1977).

8. Jim Duncan, "Place," in *The Dictionary of Human Geography*, eds. Ronald J. Johnston, Derek Gregory, Geraldine Pratt, and Michael Watts (Oxford: Blackwell, 2000), 582.

9. David Harvey, *The Condition of Postmodernity* (Oxford: Blackwell, 1989).

10. Allan J. Scott, *New Industrial Spaces* (London: Pion, 1988).

11. Rob Shields, "Social Spatialization and the Built Environment: The West Edmonton Mall," *Environment and Planning D: Society and Space* 7 (1989): 147–164; Sarah Whatmore, *Hybrid Geographies: Natures Cultures Spaces* (London: Sage, 2002); Doreen Massey, *For Space* (London: Sage, 2005).

12. Lesley Head, "More than Human, More than Nature," *Griffith Review* 31, http://www.griffithreview.com/edition-31-ways-of-seeing/more-than-human-more-than-nature.

13. Katherine Gibson, "Regional Subjection and Becoming," *Environment and Planning D: Society and Space* 19 (2001): 639–667.

14. See also, Larissa Hjorth, "Locating Mobility: Practices of Co-Presence and the Persistence of the Postal Metaphor in SMS/MMS Mobile Phone Customization in Melbourne," *Fibreculture Journal* 6 (2005), http://six.fibreculture-journal.org/.

15. William Cronon, *Nature's Metropolis: Chicago and the Great West* (New York: W.W. Norton & Co., 1992), 76.

16. Rowan Wilken, "From *Stabilitas Loci* to *Mobilitas Loci*: Networked Mobility and the Transformation of Place," *Fibreculture Journal* 6 (2005), 1, http://six.fibreculturejournal.org/.

17. Morse, "An Ontology of Everyday Distraction," 195.

18. Wilken, "From *Stabilitas Loci* to *Mobilitas Loci*," 1; Gerard Goggin, *Cell Phone Culture: Mobile Technology in Everyday Life* (London: Routledge, 2006).

19. Hjorth, "Locating Mobility," 1.

20. For example, see, David Morley, "What's 'Home' Got to Do with It?: Contradictory Dynamics in the Domestication of Technology and the Dislocation of Domesticity," *European Journal of Cultural Studies* 6 (2003): 435–458.

21. Chris Gibson, "Digital Divides in New South Wales: A Research Note on Socio-Spatial Inequality Using 2001 Census Data on Computer and Internet Technology," *Australian Geographer* 34 (2003): 239–257.

22. Cf. Ingrid Richardson, "Mobile Technosoma: Some Phenomenological Reflections on Itinerant Media Devices," *Fibreculture Journal* 6 (2005), http://six.fibreculturejournal.org/

23. Philip O'Neill, "Infrastructure Financing and Operation in the Contemporary City," *Geographical Research* 48 (2010): 3–12.
24. Gerard Goggin and Christina Spurgeon, "Premium Rate Culture: The New Business of Mobile Interactivity," *New Media & Society* 9 (2007): 753–770.
25. Gerard Goggin, *Cell Phone Culture: Mobile Technology in Everyday Life* (London: Routledge, 2006).
26. Rickie Sanders, "The Triumph of Geography," *Progress in Human Geography* 32 (2008), 179–182.
27. Karen Piper, *Cartographic Fictions: Maps, Race and Identity* (New Brunswick, NJ: Rutgers University Press, 2002).
28. Jeremy Black, *Maps and Politics* (London: Reaktion, 1997).
29. Sanders, 181.
30. Chris Gibson, Susan Luckman, and Julie Willoughby-Smith, "Creativity without Borders? Re-thinking Remoteness and Proximity," *Australian Geographer* 41 (2010): 25–38.
31. All interviews from which quotes are drawn in this chapter were conducted in 2007 and 2008 in Darwin, by Julie Willoughby-Smith and Susan Luckman. In particular, we acknowledge here Julie's invaluable assistance. For full breakdown of interview participants and methodology, see: Susan Luckman, Julie Willoughby-Smith, and Chris Brennan-Horley, *Ethnographic Interviews: Progress Report*, report prepared for Darwin City Council, NRETA Arts and Museums Division and Tourism, Northern Territory (Adelaide: University of South Australia, 2008).
32. For Telstra's interactive coverage map, refer to http://www.telstra.com.au/mobile/networks/coverage/state.html.
33. Areas beyond mobile reception range are covered instead by Telstra's satellite phone service. However, satellite phone handsets do not share the same functionality in terms of internet accessibility as their mobile counterparts, focussed instead on voice and basic data transmission. (A dial-up internet service with maximum transmission speeds of up to 2.4Kbps—similar to a dial-up modem—is offered through the satellite service.) Connection to a PC is essential to browse the internet, rather than directly via the handset. Despite Telstra's mobile coverage maps—see "Telstra mobile network coverage" Telstra Corporation, accessed 28 October 2010, http://www.telstra.com.au/mobile/networks/coverage/state.html—indicating an equal level of connection via satellite, these services remain relative to the number of satellites in orbit at any particular time. The geographic position of the handset to its immediate geographic surrounds also affects connection, with signals weakening in areas with thick canopy cover, between tall buildings, or in narrow gorges. See "Telstra satellite network information," Telstra Corporation, accessed 15 October 2010, http://www.telstra.com.au/mobile/networks/network_info/satellite.cfm
34. Romina Ricciardi, *Inequities in Telecommunications: A Case Study in Regional Mobile Telephone Access*, BSc Honours Thesis (Sydney: University of NSW, 2000).
35. "Telstra Mobile Network Coverage" Telstra Corporation, accessed 28 October 2010, http://www.telstra.com.au/mobile/networks/coverage/state.html.
36. Pierre Bourdieu, *Distinction: A Social Critique of the Judgement of Taste*, trans. Richard Nice (New York: Routledge, 2004).
37. Michel de Certeau, *The Practice of Everyday Life*, trans. Steven Rendall (Berkeley: University of California Press, 1984).
38. Kate Bowles, "Rural Cultural Research: Notes from a Small Country Town," *Australian Humanities Review* 45 (2008): 83–96.

39. Committee on Darwin and Neville Wran, *Committee on Darwin Report* (Canberra: Committee on Darwin Secretariat, 1995), 27.
40. Department of Communications and the Arts, *Creative Nation: Commonwealth Cultural Policy* (Canberra: Department of Communications and the Arts, 1994).

8 Still Mobile
A Case Study on Mobility, Home, and Being Away in Shanghai

Larissa Hjorth

INTRODUCTION

With the rise of networked, mobile technologies no longer just the preoccupation of developed countries, we are witnessing a growing diversity of instances in which place matters. In each location, what constitutes mobility (and immobility) is different. These contingencies are shaped by technological, socio-cultural, linguistic, and governmental factors, to name a few. Far from eroding a sense of home and kinship ties, mobile technologies are reinforcing the multiplicities of what constitute a sense of place. As social geographer Doreen Massey observes, a sense of place is more than a physical, geographic experience.[1] Indeed mobile, networked technologies—that is, mobile media—not only transform how we experience place in everyday life, they also highlight that place is more than just physical geography. Thus, despite the burgeoning of numerous forms of mobility—geographic, technological, socioeconomic, and physical—as part of global forces, mobile media are helping to facilitate the significance of place.[2] That is, mobile media contribute to an understanding of place as not only geographic and physical but also as evoking cartographies of the imaginary, emotional, mnemonic, and psychological.

As an extension of domesticated technologies (such as television and radio) and their attendant "mobile privatization,"[3] mobile technologies encapsulate some of the paradoxes of what it means to experience place today. As notions of mobility continue to gather speed and complexity across a variety of realms—geographic, socioeconomic, technological—it is the mobile phone that has often been the lightening rod for debates around these shifts. Underscoring Raymond Williams's "mobile privatization" is the re-articulation of domesticity beyond a simple physical place; domesticated technologies from the television onward have sought to blur boundaries between the lived/imagined, global/local, and inside/outside. In particular, mobile technologies have been important vehicles in extending co-present practices of place. Place can be lived and imagined while both being here and there, home and away. This co-present sense of place is encapsulated by the role of domesticated technologies in configuring a sense of home.[4]

While domestication may have moved out of the home—literally, in the case of the mobile phone—emotional and psychological ties to the home as a symbol of family, belonging and identity continue to grow. In this way, physical and geographic movements away from home—that can be associated with dislocation and loneliness—can be ameliorated ("put on hold") by the deployment of co-presence facilitating domestic technologies like mobile phones. For Mizuko Ito, mobile technologies allow users to reconnect to a sense of place—emotionally, electronically, psychologically—while physically being on the move.[5] In her study of Filipino workers abroad in Hong Kong, social geographer Deirdre McKay observes how mobile technologies allow these women more freedom to keep in constant contact with their family back home.[6] Like other domesticated technologies,[7] the processes involved are far from simple or finalized, as each specific site locates and adapts to new cultural artifacts in a series of exchanges. We domesticate domesticated technologies such as TV and phone as much as they domesticate us.[8]

As David Morley observes, domestic technologies such as the mobile phone help to relocate domestic ties elsewhere.[9] Mobile technologies allow us to be tethered to the home—we are free to roam geographically from the physical space of the home, but the mobile phone helps us to remain constantly connected. But this electronic anchor is not without a series of paradoxes: the mobile phone sets us "free" to roam geographically, but its always-on function means that we are seemingly always available. This "wireless leash,"[10] or "free but on a leash,"[11] syndrome creates new types of outsourcing and full-timeness of domestic labor as noted by Misa Matsuda's notion of the mobile phone in Japan (*keitai*) as "mum in the pocket."[12] While previous domestic technologies like the TV functioned within the physical space of home to reinforce psychological notions of domesticity, mobile technologies allow these relocated domestic ties to be played out outside the physical space of the home. In the case of diaspora, mobile technologies have helped to keep family connections strong despite physical dislocation.[13]

Domesticating technologies may be underscored by new modes of "mobile privatization," but they are also fraught with feelings of disjuncture as one rides the practices of co-presence integral to the relationship between place and mobility—actual and electronic—found today.[14] The "domestic" may no longer be physically located in the actual home. This home and away paradox, in turn, shifts how we imagine and define place as a patchwork of the lived and dreamed. As the various forms of mobility take hold as part of the contemporary modern condition of blurring work/leisure distinctions,[15] how we experience and define place needs clarification. For Massey, the elusiveness of place can be best understood as a collection of "stories-so-far,"[16] and she implores us to think about places beyond places. While maps give us one sense of place, they are incomplete in conveying the complex and often competing cartographies of space. As Massey notes,

> One way of seeing "places" is as on the surface of maps . . . But to escape from an imagination of space as surface is to abandon also that view of place. If space is rather a simultaneity of stories-so-far, then places are collections of those stories, articulations within the wider power-geometries of space. Their character will be a product of these intersections within that wider setting, and of what is made of them . . . And, too, of the non-meetings-up, the disconnections and the relations not established, the exclusions. All this contributes to the specificity of place.[17]

A parallel can be drawn between this definition of place as an ongoing and perpetually incomplete process and Igor Kopytoff's[18] idea of material objects as having lives and biographies. Places and objects become a collection of stories that are rendered by people into narratives of self in order to make sense of the world. Mobile technologies are devices that harness so many ways of seeing and imagining place—from new co-present visualities (GPS, camera phones, social media), to co-present forms of intimacy through telephony and texting. As such, mobile media can be seen to highlight the ways in which subjectivities and context are massaging (or domesticating) global technologies. In a flux of numerous mobilities, mobile media practices provide insight into the complex and paradoxical ways we imagine, sketch, and redraw place, while being both home and away.[19]

One such place that is witnessing dramatic shifts in its usage of mobile technologies—alongside changes in various forms of mobility—is China. Over the last ten years, cities such as Shanghai have not only seen the implementation of new technologies, such as Internet and mobile phones, in educational and work settings, but these changes have accompanied a trend of studying away from home by the *ba ling hou* generation. Born in the 1980s, this is the first generation to grow up with the Internet and new media as part of everyday life. They are also a generation that has, thanks to the ubiquity of information and communication technologies (ICTs), and its attendant technological and thus geographic mobility, been allowed to move physically to study—a freedom unheard of for previous generations.

Often from one-child families, these *ba ling hou* are incredibly close to their parents and, although physical, geographic mobility for education is a given for this generation, there is a need for continual contact with their friends and family back home. In studying away from home, the deployment of mobile media—especially through social networked sites (SNS) like QQ and Renren—has been integral for this generation as they negotiate home and away. Technologies such as mobile media allow this generation to move geographically from the home and yet still remain connected, tethered to a sense of home as an emotional and psychological space of belonging.

Through the constant contact afforded by mobile connectivity, the *ba ling hou* are free to roam with their "parents-in-the-pocket" via the mobile phone. In this way, mobile and social media create a relocation of these domestic ties outside the physical space of the home. By exploring the different techniques that are used to facilitate this, and associated feelings, we can gain a sense of how the home can be perpetually experienced—as both lived and imagined—while physically away. This phenomenon demonstrates a very particular version of place as the stories-so-far and the ways in which mobile technologies help narrate specific cartographies of place and home associated with family and history. As this study identifies, the sketching of the cartographies of place is geographic and physical and emotional and social.

In this chapter, I focus upon students who migrate, often from small towns in China, to Shanghai for university-level education. In this physical migration, the deployment of mobile media has been pivotal in these students' maintenance of various aforementioned notions of mobility. As the first generation of media literates to benefit from an educational system in which new technologies were vital—online computer networking was provided by the China Education and Research Network (CERNET) implemented from 1994, and the Electronic Information Service System (EISS) in 2000[20]—these *ba ling hou* are often teaching their parents and grandparents how to use these technologies in order to maintain contact and connection. For many of these students, who are predominantly from working class backgrounds, the socioeconomic mobility afforded to them through education is being transferred in unofficial ways via cross-generational media literacy. This media literacy illustrates innovative use—"making-do"—of available technologies and social media like QQ (the biggest and longest running social media in China).

This chapter focuses on a case study of 80 Fudan University students, which I conducted in Shanghai during June and July 2009 (with 40 students) and again in June and July of 2010 (with another 40 students). Through focus groups and one-on-one, in-depth interviews with both students and their parents, I explore some of the ways in which these students traverse home and away through mobile and Internet technologies. We will see how the different modes of co-presence and cross-generational media literacy operate to maintain a tether to the home: as an emotional, psychological, social, imagined, embodied, and actual place. Here we can think about the complexity of what constitutes home as something that is a "chiaroscuro of setting, landscape, ritual, routine, other people, personal experiences, care"[21] that is reflected in, and by, co-present technologies such as social and mobile media. In Shanghai, students often use mobile media to communicate with their parents—from voice calls, to SNS such as QQ. In the face of various forms of mobility experienced by these students, I look at some of the ways in which mobile media are deployed to reinforce a sense of place, past, and locality. To do so, I begin outlining the technoculture that is China—and specifically Shanghai—and the role played by mobile and social media. I then turn to the case study of how students and parents are using and conceptualizing the role of mobile media in their negotiation of home and away.

LOGGING ONTO THE SOCIAL: CONTEXTUALIZING CHINA AND SOCIAL MEDIA

Across the plethora of different SNS, modes of access (via PC or mobile), and online communities in the region, there are commonalities as well as distinctions that mark the Asia-Pacific's embrace of Web 2.0. For youth across the region, SNS is not only a fundamental part of everyday life and social capital, it is also a space that allows them to maintain cross-generational contact when geographic distance might be involved. It is no longer just "youth" that are using SNS as adult on adult and cross-generational forms of dialogue and literacy expand with the increased net accessibility. Moreover, the demographics are shifting, too, as Internet access becomes not only a middle class prerogative but an integral part of the new mobile working class.[22] For example, in China, it is the working class that is growing exponentially as the main users of the Internet predominantly via the mobile phone.[23]

In particular, the Internet in China has continued to gain much global focus. The unique features of China within the global chains of ICT production and consumption over the last couple of decades[24] have been most evident within the twenty-first century through the deployment of the Internet as a site for contestation and technonationalism.[25] However, much of the literature on China's Internet has focused on its role as a contested site between public opinion and government policy. Like South Korea, the Internet in China has provided a way in which to conceptualize struggles for democracy with social media such as blogs attracting much focus,[26] while seemingly "less" political media such as SNS have been relatively overlooked,[27] despite often being deemed threatening and thus banned by the government (in 2009, Facebook and Twitter were barred).

By 2009, Internet usage in China had surpassed the global average with over 298 million users (279 million of these being broadband users) with China becoming the number one global user of the Internet.[28] This phenomenon has continued with China's voracious use of social media, like blogs, such that they now have more users than any other nation. So, while technological infrastructure and access to the Internet are still predominantly urban phenomena in China, the number of rural Internet users has reached 85 million.[29] While the Internet is mainly accessed via personal computer in urban city areas by the middle and upper class, the hundred million odd migrant working class—along with the rest of China's predominantly working class demography—deploy GPRS (General Packet Radio Service) for Internet access via the mobile phone. One of the key forms of SNS accessed by the mobile phone is QQ.

China's SNS technoculture in Shanghai features numerous examples both local and global. Local SNS include QQ, Kaixin, Renren, and Fetion; while MySpace and MSN (Facebook and Twitter were banned in 2009, although some still use it) feature as global examples. However, the use of both local and global SNS is only apparent in "cosmopolitan" cities such as Shanghai. These media practices are indicative of particular urban lifestyles that are not replicated in rural areas. For the general population of China, few local SNS are used, or even known, apart from QQ. Part of QQ's success has been ensured by its inclusion and convergence of various platforms from chatting and SNS (QQzone) to online, localized news. Via Internet portals its service is free, and its various platforms such as QQ chat can be adapted to devices such as mobile phones. This is important given that while most of the population outside of the big towns and cities do not have the Internet they all have a mobile phone. QQ represents the largest cluster of SNS services in China, resulting in QQ being among the largest in the world. Moreover, as we shall see in the case study, QQ represented a type of Chinese nationalism and many expressed a strong sense of sentimentalism toward QQ as their first point of entry into SNS. Despite the significance of QQ in emphasizing the vernacular and local of Chinese sociality—and its history as the first SNS and instant messaging (IM) program that ushered a new generation of digital natives through to Web 2.0—studies of QQ have been limited.

As Koch et al. note, "the emergence of this local homegrown QQ software indicates that it plays an important role in Chinese society."[30] Looking at QQ can help us to "understand the influence of this specific Internet platform as situated within Chinese culture."[31] And yet, as Koch et al. note, the various services and platforms of QQ mean that it differs greatly from the "Internet" in China in general. Rather, it is the portal between media, between platforms, between generations, between media experiences, between classes. QQ offers mobility at the same time as it emphasizes a sense of place. This sense of place is both geographic and physical as it is emotional and social. QQ embodies a Chinese sense of locality in the twenty-first century.

Specifically, QQ occupies a particular role in the technocultural imaginary of China. Here we see how social and mobile media can actively add to the imaginary dimensions of a sense of place and belonging. QQ is not only one of the oldest SNS in China—and thus, for many, their first SNS and introduction to social media—it is also a SNS predominantly deployed by the lower socioeconomic and rural/small town users. QQ's various platforms, such as chat, allow families to keep in contact regularly for little or no charge. These features of both convenience and lack of costs are but two in the many reasons why QQ has attracted over 376 million registered accounts.[32] One of the dominant reasons can be heard in the voices of users in which QQ is viewed as Chinese *before* it is technological.[33] That is, QQ represents China's particular nationalist feelings toward the Internet which no other SNS produces.

Moreover, QQ is emblematic of the new breed of young media literates who are transferring media knowledge to their parents and grandparents in order to maintain kinship ties while living away from home. Unlike the other

SNS (Renren and Kaixin) that are mainly for and by university students, QQ's lineage and attendant demography occupies a specific role within the lives of the young as they migrate socioeconomically through educational and IT policies.[34] QQ's particular material conditions (as a free and convenient service that can be accessed through the phone) and symbolic conditions (conjuring feelings of nationalism and nostalgia as China's oldest SNS) provide a pathway for these young media literates to keep in contact with family while studying away from home—and also to afford older generations of users a discursive, nonofficial route into new media literacy. Often, parents and grandparents begin using QQ to keep in contact with their child studying away from home, and, before long, they have developed broader skills with online technologies, including participating in gaming and informational services.[35] While younger generations may have more time, interest and accessibility (via university) to various SNS, QQ is the one common language understood *between* the generations and classes in different places in China. This factor is notable in the sample study.

In the twenty-first century, China has gained much focus as it shifts from a developing to a developed economy—a transformation in which ICTs play a pivotal role. Evidence of a burgeoning middle class in China, and their associated material symbols of success, is amplified in the context of Shanghai.[36] But, behind these images of a new middle class and progressive mobility (social, class, economic), Shanghai has another side which is emblematic of bigger technocultural and socioeconomic factors and reforms in mainland China. That is, the rise of the migrant worker and the increased deployment of mobile technologies in this transformation.[37]

While the Shanghanese middle class enjoy the luxuries of Internet enabled computers at home and on the move, in addition to mobile devices that often have 3G capacities, for the working class migrants their mobile phone is *the* access to both their new working life in Shanghai and their family and friends back home.[38] For many working class and students, 3G mobile Internet, despite the hype,[39] is not a reality. Rather, through the deployment of various methods, including GPRS, WiFi and PCs, these users are demonstrating new forms of making-do practices indicative of the growing demographies of "have-less."[40] These making-do practices of the have-less are amplified when exploring Shanghai's undulating and ever-changing technocultures and attendant forms of mobility. For many university students, Shanghai is not their home but rather the place they have come to in order to study—often leaving family and friends back home. Through the coordination of various social media networks, they are able to both enhance existing relations and also develop new ones with fellow students and part time work colleagues. In this ensemble of various SNS and IM, it is QQ that remains the enduring network for maintaining familial and old friendships.[41]

In this changing social fabric, in which various forms of mobility (geographic, cultural, and socioeconomic) can be witnessed, the significance of networked, mobile technologies, such as QQ, in helping ensure kinship ties, cannot be underestimated. Far from new media distancing them from their past and socioeconomic background, they provide a material and symbolic portal between home and away. Thus, through the lens of such mobile technologies, we can chart this transitory period in which China's technocultures and the emergent forms of mobility—especially around the new breed of youth (*ba ling hou*) that have accompanied this rise. Through networked, mobile technologies one can examine a variety of mobilities and the way a sense of place (both psychologically and physically) informs these practices.

CASE STUDY OF STUDENTS (AND THEIR PARENTS) STUDYING IN SHANGHAI

> "If I was a fish then QQ would be my water." (Bao, 25 year old, male postgraduate student)

In June and July of 2009, I conducted fieldwork that focused on university students' usage of mobile and social media. Through a variety of methods including survey, focus groups, and one-on-one interviews, I made two interesting discoveries. Firstly, a majority of my interviewees were not from Shanghai and came from lower socioeconomic backgrounds. Secondly, many spoke at length about the deployment of mobile and social media to negotiate home and away. Often they had different SNS for friends at home, as opposed to university friends and contacts in Shanghai. While the difference in media literacy across the generations was noted—many observed that their parents viewed the Internet as just another type of TV—I will focus upon the role of place and mobility served by these new media practices.

This *ba ling hou* generation have a particular narrative of place in which technologies have played a central role. As the first generation to grow up in China's emerging net culture, the *ba ling hou* are a product of the first education IT initiative that began in 1994. They are not only the generation to profit from growing up in tandem with China's Internet and technology education reforms, they are, interrelatedly, the first generation to collectively undertake tertiary study away from home, travelling either inside and outside China for their studies. As children from generally one-child families, they tend to be incredibly close to their parents and grandparents, a feeling that is exacerbated when they physically move away from home. In this physical relocation, social and mobile media play a key role in their emotional identification with the imaginary "home" and a sense of place and family.

Through social media, they not only can keep in perpetual contact with their family via their phones, they are also able to collect, archive and memorialize their "stories-so-far," to construct the patchwork that is a sense of place. Memories, stories, pictures, and exchanges become paradoxically *ephemerally memorialized* within, and around, the vehicles of social mobile media. The fleeting and transitory nature of the digital perfectly matches the ephemeral everyday exchanges that circumnavigate both the phatic and the poetic. These cartographies of the vernacular and everyday are captured by social and mobile media, operating as contemporary versions of older co-present forms of intimacy, such as the postcard.[42] However, new media also produce new types of intimacy and re-imaginations of place.

The need to maintain sometimes daily contact with their parents back home had required the students to use a variety of mobile and social media; often teaching their parents about new media, like QQ's video service (similar to Skype), which provide cheap (in most cases free) and convenient services. In many cases, students claimed to have taught their parents to use such media so that they could maintain communicative exchange and connection despite being in different physical places. Prior to discussing some examples, I will outline the educational policies that accompanied the rise of the *ba ling hou* generation.

For the average *ba ling hou* student, they are part of the EISS generation, growing up with the benefits of these policies informing their high media literacy whereby the Internet is viewed as a pivotal part of everyday life. The EISS, along with the general rise of Web 2.0 and its participatory entertainment industries, has ensured that for most university students in China today, the Internet and mobile technologies are an integral part of their daily life. Through this media literacy, along with China's growing economic prosperity, many of the *ba ling hou* generation can travel—often for study. This link between new media policy reforms, growing access to new media, and also high new media literacy, has allowed the *ba ling hou* to negotiate a sense of mobility and place unimaginable for previous generations. Current Fudan University students are a good example of these new media literates and their attendant forms of mobility and co-presence.

One female postgraduate student (aged 26) spoke of her time at Fudan University as a period marked by shifts in technological access and availability. She noted in the first year (2004) there were few computers and Internet portals available in classrooms, however, within a matter of two years she observed the overt rise in computers and accessibility—to the point where now everyone has a computer and Internet access is all-pervasive. Indeed, at the end of the EISS project, the impact is immeasurable, especially for the current young generation of university students for whom the Internet is a pivotal part of their everyday life, and imagining one day without it would be more than difficult and inconvenient, it would be unthinkable. It is this new generation of digital literates that is helping to encourage older generations, as well as being the teachers for the working class, to embrace the making-do mobile Internet technocultures of China tomorrow.

Given that all but one of the focus group participants was from elsewhere than Shanghai, issues around family, home, and how SNS functioned to facilitate this within the various forms of mobility (class, technological, geographic, social), dominated the discussion. For one female postgraduate student aged 24, one's origin strongly influenced what media they used. She noted changes of SNS dependent on life stages (work, etc.) and where one is living; she noted that QQ continued to play a special role in maintaining her familial and kinship ties, and, for this reason, she had taught her family to use QQ. This was a common pattern found in almost all of the participants—regardless of different SNS and mobile media usage: QQ continued to function to maintain kinship ties and geo-social and economic connections with their past. One female respondent, aged 27, commented:

> I think your relationship with technological communication tools depends where you are in China and where you are from . . . After I finished my Bachelor degree at Taijin University I came to Shanghai (Fudan University). I used QQ a lot when I was in university and then I started using MSN when I was in senior year because I was doing an internship. In the office, MSN is really important to transfer documents . . . I remember in the summer of 2004 when Xiaonei (now Renren) had been launched they would give you a free ice cream or chicken leg if you joined. Now it's really popular, almost everyone has a Xiaonei account. In the beginning no one knew of it so it was a good way to get people to know quickly. The more I used it the more interesting it became. At first, I used it to find my elementary school friends and middle school and high school friends—we are in different places but Xiaonei helps us to reconnect and stay in contact. For me, Xiaonei is not a place to make new friends, just to reconnect to already known friends and contacts . . . But I continue to use QQ for my parents. Last year they didn't know how to use a computer, let alone QQ, so I showed them. Now my family has moved to a new place so we use QQ video so we can see each other whilst chatting.

As this respondent noted, where one comes from plays an important part in their uptake, and perceived relevance, of new media. And this difference isn't just a mere rural versus urban divide. In a country as big as China and with a history so long and enduring, the quick embrace of twenty-first media practice has taken on various mutations subject to location. Although the respondent here is talking about a physical, geographic sense of place, what she is alluding to is that those places are also imbued with emotional and social particularities that translate in the use and adaptation of the technologies.

When asked to discuss how migration had affected their usage of SNS and mobile media, the focus group discussed the various SNS and the types of deployment and associated social networks. Many were in agreement about the different usages of MSN and QQ and the attendant class and lifestyle images. Issues such as class, place, technological access, and language, arose as important factors ensuring QQ's ongoing success and relevance for both young and old generations entering Web 2.0 mediascapes.

For many, QQ offered the most options through its various localized services. This service is not only a tailoring to a physical, geographic sense of place through GPS, it also highlights that each geographic location has its own stories and narratives that imbue certain issues with more emotion and meaning. The news reports not only have stories about events that happened in that location but also customize the content for what that place tends to value and find of interest. Here we see that place is both geographic and physical as it is emotional and social. As respondents talk about localizing qualities we are, at the same time, given an image of the elusiveness of place—its collection of stories are both official, or part of "wider power-geometries of space,"[43] but also unofficial and vernacular. Social mobile media particularly afford a space for the latter, allowing small and informal micronarratives to blossom with their emphasis upon the creative vernacular.[44] For one female respondent, aged 26,

> I noticed when I came to Shanghai people were using MSN and Skype which I hadn't seen used elsewhere . . . In other parts of China there are a lot of people who don't have a computer in their home, let alone Internet. So QQ is much more popular with these people as it can be accessed via the phone mobile—something that everyone has. MSN is popular here in Shanghai, unlike the rest of China. A lot of children who don't have a computer or Internet access at home go to an Internet café. At those places, they don't have MSN downloaded onto the desktop and it requires people to be able to use English language, so many just use the already available QQ. Young students, especially college and elementary, use QQ, and in developing areas there would be more people using QQ than anything else. But, in the big cities like Shanghai and Beijing, there are so many companies and so many white collars so they use MSN a lot because they are older.

For another respondent (a 22-year-old female) the usage of QQ every day provided a familiar experience in an unfamiliar environment (as she moved away from home to attend university). It also helped to maintain her contact with her family, alleviating her feelings of homesickness:

> I use QQ almost every day. I use both my computer and mobile phone all day, every morning and every night. I use various services of QQ. This is the first year I have lived away from home and so I use QQ mainly to keep in contact with my family at home as I board away from home. I use QQ almost every day to make contact with my family. I get very homesick and so QQ helps with that. Last month my mother had an operation and I couldn't come home because of exams so instead I used QQ to keep in contact with her continuously.

Here, the above respondent is highlighting that what constitutes a sense of place and home is marked by the emotional and social. It is an affective cartography that is marked and maintained by the ephemeral, personal and micronarratives of social mobile media. QQ doesn't just connect the respondent with a physical place, it allows her to be co-present and to negotiate feelings of homesickness. Rather than replacing physical contact, media such as QQ allow for the small, informal, micro exchanges, like those experienced when one is in intimate proximity. Through the feelings of homesickness, in which there is movement between a state of inertia, distress and sadness, as a result of actual and perceived separation with a specific home environment or attached objects, we can gain some sense of how place operates upon various levels—psychological, emotional, and social, to name a few. Sometimes we see that media, such as social media and mobile technologies, help alleviate homesickness, whereas for others, such mediation operates to only further highlight the actual distance and difference.[45]

In the 2010 fieldwork, QQ and Renren still played a major role in mobile media communication. For one family from the Fujian Province—in this case, a father, mother, and daughter—the deployment of mobile media in maintaining contact while separated physically by geography was pivotal. Interestingly, the daughter used different methods when talking to her father and mother; for her father, she used voice call, whereas she used QQ with her mother. The father she spoke to once a week, the mother every second day. The chat function on QQ was enjoyed by mother and daughter, although the daughter noted a common problem was that her mother typed too slowly. The mother mainly used the Internet for communication, especially with her daughter. For the father, "the cell phone was important and yet annoying." Working for the government where the landline was used predominantly, he used the mobile phone for family and friends. When asked about the difference of usage between rural and urban users, the mother noted:

> I think the Internet is popular in Chinese family in the cities, so maybe every family has Internet access. And except communicating with my daughter, I have a few, not too many, good friends that I will communicate with on the QQ.

Here we see that the importance of new media is not just dependent upon generational and lifestyle factors, but can also be understood in terms of the rural and urban divide found in many countries. She continued, "my daughter will do a lot of things on the Internet besides communicating with friends, such as searching for information and entertainment. I think she is in many school online communities. But the priority is for work." For the daughter, the perceived differences were summed up as follows: "I think most of them [her parents' generation] use mobile phones, and they send messages a lot, while we maybe use mobile at work, we use QQ on mobile. And we use emails a lot."

For another family, a father and daughter from the Jiangsu Province, the main difference was around the agency of the user that was enabled by new media. For the daughter, the father viewed the Internet as a source of entertainment, much like a twenty-first century version of the TV. Whereas, for the daughter, the Internet was mainly for socializing, especially for maintaining social capital between past and present contacts. Like most of the parents, the only time they came to Shanghai was to visit their child during their holidays. When I asked the father how he kept in contact with his daughter he said: "I use the Internet to watch movies and to read the news reports, not to contact my daughter. I use the cell phone to contact her." When I asked whether he viewed the Internet at home as a twenty-first century TV, he replied, "It's much better than TV!"

Many of the interviewed students and the parents talked about the important role new media played in maintaining contact when physical distance was imposed. While they noted that these forms of mediated communication would never replace actual contact, the parents noted the significance of this media for the *ba ling hou* generation. Mobile media has accompanied the *ba ling hou's* mobility as they traverse both physical places and also socioeconomic terrain. In the face of these various forms of mobility (class, geography, social, and economic) in China's transforming technoculture, the pivotal role provided by mobile media remained constant contact. It connects generations, migrating across geographic distance, as well as securing and locating a sense of place, family, and community. Whether it be a source for gaming, TV, or socializing, mobile media is playing an important role in cross-generational communication practices as they negotiate many forms of mobility and a sense of place.

KEEPING IT IN THE FAMILY: CHINA'S MOBILE TECHNOCULTURES

As the young move away from home in the quest for a university education, often there will be many feelings of homesickness. The social mobility of a university education has also given way to other forms of mobility thanks to the implementation and rise of new media in China. With mobile and social media, this new generation's educational rite of passage sees studying away from home as an imperative. Social and mobile media afford this generation new ways to remain electronically (and psychologically) tethered to the home and yet geographically away from the home.

As the first generation to benefit from policies such as EISS, implemented in 2000, the *ba ling hou* are using mobile and social media as essential tools in negotiating place and co-presence as they traverse being home and away. For the youth studying in another place, mobile media provide efficient and convenient ways to keep connected with family and friends at home, and help to alleviate some of the feelings of loneliness and sadness. For these students, social networked media are an integral part of their everyday life.

This case study of university students and their parents in Shanghai is the first of many that need to be conducted if we are to understand the relationship between these emerging forms of mobility and China's twenty-first century technoculture. As the first generation to grow up with the Internet, this generation—and their relationship to other generations—can provide some new models for media literacy. Moreover, while these new media literates are demonstrating a new form of mobility across geographic and socioeconomic differences, they are highlighting the importance of maintaining a connection with their past and a sense of place. Here we are witnessing only part of the complex picture that is place as an imagined and actual, emotional and social, geographic and physical concept. Through the micronarratives and ritualized practices of social mobile media usage, we can gain insight into some of the ways place—as a collection of stories-so-far—is negotiated. This media usage not only illustrates how intimacy has always been mediated—that is, through language, gestures, and memory, but also how different types of media, used by different generations in different places, can bring a multiplicity of interpretations to what place and home means. We can see some of the many ways in which social mobile media have become a space for navigating numerous types of place—both actual and imagined—to highlight that in an age of mobile technologies the types of maps we need of place are less of the geographic, physical, or GPS sort, and more like cartographies of the emotional. As a lens through which to view emerging forms of intimacy, kinship, and media literacy, mobile media demonstrates that despite the growing modes of mobility— geographic, technological, socioeconomic—home (or a sense of it) is never far away.

NOTES

1. Doreen Massey, "Questions of Locality," *Geography* 78 (1993): 142–9.
2. Larissa Hjorth, "Locating Mobility: Practices of Co-presence and the Persistence of the Postal metaphor in SMS/MMS Mobile Phone Customization in Melbourne," *Fibreculture Journal* 6 (2005), http://six.fibreculturejournal.org/.
3. Raymond Williams, *Television: Technology and Cultural Form* (London: Fontana, 1974); David Morley, "What's 'Home' Got to Do with It?," *European Journal of Cultural Studies* 6 (2003): 435–458.
4. Morley, "What's 'Home' Got to Do with It?"
5. Mizuko Ito, "Mobiles and the Appropriation of Place," *Receiver Magazine* 8 (2002), http://www.receiver.vodafone.com/
6. Deidre McKay, "'Sending Dollars Shows Feeling'—Emotions and Economies in Filipino Migration," *Mobilities* 2 (2007): 175–194.

7. Morley, "What's 'Home' Got to Do with It?"
8. Roger Silverstone and Leslie Haddon, "Design and Domestication of Information and Communication Technologies: Technical Change and Everyday Life," in *Communication by Design: The Politics of Information and Communication Technologies*, eds. Roger Silverstone and Richard Mansell (Oxford, UK: Oxford University Press, 1996), 44–74
9. Morley, "What's 'Home' Got to Do with It?"
10. Jack Qiu, "Wireless Working-Class ICTs and the Chinese Informational City," *Journal of Urban Technology* 3 (2008): 57–77.
11. Michael Arnold, "On the Phenomenology of Technology: The 'Janus-Faces' of Mobile Phones," *Information and Organization* 13 (2003): 231–256.
12. Misa Matsuda, "Mobile Media and the Transformation of the Family," in *Mobile Technologies: From Telecommunications to Media*, eds. Gerard Goggin and Larissa Hjorth (New York: Routledge, 2009), 62–72.
13. McKay, "'Sending Dollars Shows Feeling'."
14. John Urry, "Mobility and Proximity," *Sociology* 36 (2002): 255–274; Ito "Mobiles and the Appropriation of Place;" Margaret Morse, *Virtualities: Television, Media Art, and Cyberculture* (Bloomington: Indiana University Press, 1998).
15. Judy Wajcman, Michael Bittman, and Jude Brown, "Intimate Connections: The Impact of the Mobile Phone on Work/Life Boundaries," in *Mobile Technologies: From Telecommunications to Media*, eds. Gerard Goggin and Larissa Hjorth (New York: Routledge, 2009), 9–22.
16. Doreen Massey, *For Space* (London: Sage, 2005).
17. Massey, *For Space*, 130.
18. Igor Kopytoff, "The Cultural Biography of Things: Commoditization as Process," in *The Social Life of Things: Commodities in Cultural Perspective*, ed. Arjun Appadurai (Cambridge, UK: Cambridge University Press, 1986), 66.
19. Hjorth, "Locating Mobility."
20. As the first generation to grow up in China's emerging net culture, the *ba ling hou* are a product of the first education IT initiative that began in 1994. This initiative, in the form of the first national network, was called CERNET (Chinese Education Research Network). Many of the *ba ling hou* were the first group of students to be affected by the EISS (an acronym for 'Electronic Information Service System' or in Chinese 'xiaoxiaotong') policies in which the government orchestrated, over a ten-year period (2000–2010), the implementation of computers and online education within primary and secondary schools. The EISS was initiated in November 2000, as a project spanning ten years to enable 90 percent of the independent middle schools and primary schools throughout the country to have access to the Internet, along with increasing the deployment of online resources to be shared amongst teachers and students.
21. Edward Relph, *Place and Placelessness* (London: Pion, 1976) 29.
22. Qiu, "Wireless Working-Class ICTs."
23. CNNIC [China Internet Network Information Center] (2009) *Statistical Survey Report on the Internet Development in China* (January 2009), http://www.cnnic.cn/uploadfiles/pdf/2009/3/23/131303.pdf.
24. Richard Robison and David S. G. Goodman, eds., *The New Rich in Asia* (London: Routledge, 1996); Stephanie H. Donald, Michael Keane, and Yin Hong, eds., *Media in China: Consumption, Content and Crisis* (London: Routledge, 2002).
25. Jens Damn and Simona Thomas, eds., *Chinese Cyberspaces: Technological Changes and Political Effects* (London and New York: Routledge, 2002); Zixue Tai, *The Internet in China: Cyberspace and Civil Society* (New York: Routledge, 2006).

26. Axel Bruns and Joanne Jacobs, eds., *Uses of Blogs* (New York: Peter Lang, 2006); Geert Lovink, *Zero Comments: Blogging and Critical Internet Culture* (New York: Routledge, 2007); Haiqing Yu, "Blogging Everyday Life in Chinese Internet Culture," *Asian Studies Review* 31 (2007): 423–433.
27. Pamela Koch, Bradley J. Koch, Kun Huang, and Wei Chen, "Beauty is in the Eye of the QQ User: Instant Messaging in China," in *Internationalizing Internet Studies*, eds. Gerard Goggin and Mark McLelland (London: Routledge, 2009), 265–284; Yu, "Blogging Everyday Life in Chinese Internet Culture."
28. CNNIC *Statistical Survey* (2009).
29. CNNIC *Statistical Survey* (2009).
30. Koch, et. al., "Beauty is in the Eye of the QQ User," 265.
31. Koch, et. al., "Beauty is in the Eye of the QQ User," 265.
32. "Market Entry & Business Acceleration in China, Japan & Korea," http://www.web2asia.com/, accessed June 27, 2011.
33. Koch, et al. "Beauty is in the Eye of the QQ User."
34. CNNIC *Statistical Survey* (2009).
35. QQ, in particular, has become a form of glue for cross-generational mobility, marking a new pathway of lifestyle cultures in China. Just as migrant workers—or what have been called the "floating population"—use new mobile technologies for SNS such as QQ to reinforce social networks and kin relationships in practices called "communities of the air," the usage of QQ by university students is marking new forms of cross-generational media literacy that is helping bridge the gap as they migrate, via education and IT policies, into new lifestyles. While these university students from lower socioeconomic backgrounds coming to Shanghai cannot be defined as traditionally "working class," they do represent a small component in China's changing social fabric in which labor, capital and technologies are being immobilized—with both positive and negative outcomes. For many of these students coming from one-child families, their parents have worked hard to give them the opportunity to study at university.
36. Jun Wang and Si Yau Lau, "Gentrification and Shanghai's New Middle-class: Another Reflection on the Cultural Consumption Thesis," *Cities* 26 (2009): 57–66.
37. Ding Wei and Tian Qian, "The Mobile Hearth: A Case Study On New Media Usage and Migrant Workers' Social Relationship," paper presented at *ANZCA Conference*, Queensland University of Technology, Brisbane, 8–10 July, 2009, http://www.anzca.net/download-document/75-the-mobile-hearth-a-case-study-on-new-media-usage-and-migrant-workers-social-relationships.html.
38. Metropolises such as Shanghai are central sites for migrant labor, especially as the modernization of China through such projects as the damming of the Yangtze River force manual laborers to move. Indeed, Shanghai's role in the migration of workers can be seen as reflected in the recent relaxing of residency rules in Shanghai. However, according to a BBC report, these "new rules are only thought to benefit 3,000 of the city's estimated six million migrant workers" of which "one-third of the population are migrant workers who have come from other parts of the country." ("Shanghai Relaxes Residency Rules," *BBC News online*, http://news.bbc.co.uk/go/pr/fr/-/2/hi/asia-pacific/8107518.stm.)
39. CNNIC *Statistic Survey* (2009).
40. Qiu, "Wireless Working-Class ICTs."
41. For those of the working class, moving from small towns to large cities for work has become a reality in which mobile technologies feature

predominantly as a way to maintain ties with family and friends back home as well as helping with new work-related connections (see Wei and Qian, "The Mobile Hearth").

42. Hjorth, "Locating Mobility."
43. Massey, *For Space*, 130.
44. Jean Burgess, "'All Your Chocolate Rain Are Belong to Us?' Viral Video, YouTube and the Dynamics of Participatory Culture," in *The VideoVortex Reader*, eds. Geert Lovink, et al. (Amsterdam: Institute of Network Cultures, 2008).
45. Larissa Hjorth, "Home and Away: A Case Study of the Use of Cyworld Mini-hompy by Korean Students Studying in Australia," *Asian Studies Review* 31 (2007): 397–407.

9 Connection and Inspiration
Phenomenology, Mobile Communications, Place

Iain Sutherland

INTRODUCTION

Media and communications technologies have often been pitted against authentic place-making practices because they are conceptualized as homogenizing human experience across different places. A dominant global capitalist culture industry, enabled by global media networks, so the argument goes, undermines the uniqueness of place, whose authenticity depends on the maintenance of local culture articulated via the local interaction of a place's inhabitants.[1] Many of these critiques derive from phenomenological and broadly humanist perspectives, which emphasize the body, its movement, interactions, perceptions and habits as crucial factors in place-making and the experience of place. Not just media, but also transport technologies are implicated in this degeneration of place. This chapter explores a contrary possibility: that while certain media and transport technologies may produce a quite *different* perspective on what a place is, they nevertheless may also contribute positively to the phenomenological construction of the experience of place. This, it will be argued, is particularly the case regarding technologies which are intimately connected with those objects of enquiry central to phenomenological accounts: the body, its movement, interactions, perceptions, and habits in the unfolding of everyday life. One such set of technologies is to be found in mobile communication devices. This chapter explores the ways in which phenomenological perspectives can be made to illuminate contemporary place-making practices in the context of ubiquitous mobile communications systems and how these perspectives themselves might be reconceptualized to accommodate and account for them.

These perspectives are to be found in the work of a group of humanist geographers writing in the late 1970s—principally Yi-Fu Tuan,[2] Edward Relph,[3] and David Seamon.[4] Seamon's work in particular, provides an incisive analytical tool in the current context both because it explicitly builds on Tuan and Relph's foundational thinking and because of its empirical application of the phenomenological approach. "Environmental experience groups," consisting of American college students, were asked to reflect on their everyday experience of place in order to shed light on its often submerged, bodily, and habitual dimensions. Seamon suggests that his research participants orient themselves to place primarily via "body subject"—"the inherent capacity of the body to direct behaviors of the person intelligently, and thus function as a special kind of subject which expresses itself in a pre-conscious way, usually described by such words as 'automatic,' 'habitual,' 'involuntary' and 'mechanical'."[5] This emphasis is useful in conceptualizing the role of contemporary mobile communications systems because the latter are now so ubiquitous—and hence mundane, that they are submerged beneath, and at the same time, utterly integral to, the regimes of movement, interaction and perception that characterize everyday life in contemporary Western societies. They are, to this extent, enrolled in the phenomenological processes by which anonymous space is transformed into familiar and meaningful place.[6] Seamon's empirical commitment, as will become clear, is also adopted in the course of this chapter.

To approach mobile devices in this light, however, demands a reconfiguration of phenomenological-geographic thought, which often treats media technologies as "disembedding" because of the ways in which they variously "remove" their users from the immediate environment or cause the user to be "in two places at once."[7] This suggests that place is relatively static and bounded, that it is created through an apposite immobility of dwelling and inhabitation[8] with minimal intervention between the embodied subject and his or her world. Two polarities are delineated, each stemming from a focus on the embodied subject. One is a question of technology—the immediate and embodied being valorized, while the mediated and technical remains *a priori* antithetical to place-making practice. The other is a question of scale—the local and intimate being associated with authentic relations to place while spatial relations across larger scales produce only disjuncture. This chapter suggests that there is a third axis apparent in phenomenological relations to place which can be applied to such analyses which maintains the emphasis on embodied practice but does not foreclose the possibility that place-making can be variously technologically enabled and can unfold across different scales. In foregrounding this aspect of phenomenological experience, the key questions are not whether experience of place is technologized or not, or whether it is local or dispersed. Rather, they concern the ways in which both body and technology are enfolded into rhythmic processes of establishing an everyday familiarity with place and at the same time maintaining a degree of unexpectedness. By highlighting these rhythms, a more nuanced account of the role of technologies in formulating experience is developed.

In untangling these often complex issues, the chapter focuses on a piece of empirical work carried out in 2006. This work consisted of a series of interviews and participant observation sessions with Tristan and James, two Australians living in Barcelona, examining the ways in which mobile communications were incorporated into their everyday lives. The purpose of this work, within the context of a larger study, was to explicate those aspects of the interaction between places and people and their various media devices which, through repetition and sedimentation into habit, often sink below the threshold of accountability on which participant observation techniques rely. This threshold is an important concept throughout the chapter since it is our bodily, practical, sometimes unconscious engagement with place with which phenomenological geographers are concerned. Tristan and James's experience foregrounds an instance of partial disjuncture. It focuses attention on place-making made visible by its disjunction from the familiar settings in which phenomenological accounts of place tend to unfold. They are both well traveled, having made their way to Barcelona via India, Canada, Japan, Hong Kong, Turkey, England, Ireland, and New York, and chose Barcelona as a place to settle from a wide range of options. "Asia just wasn't very appealing. We'd traveled through Asia. The States certainly wasn't appealing," says James. The degree of choice involved here makes Tristan and James' insights particularly valuable, since their aims and desires as mobile citizens are fully articulated in both word and deed.

The present case also foregrounds another facet of the experience of place which often eludes phenomenological accounts. That is, place-making practices are rich, varied, and heterogeneous. Phenomenological geography is often marked by an attempt to define an often elusive and usually reductive essence of "being in place," which valorizes "existential insideness"[9] and "at-homeness."[10] In this respect Tristan and James's experience could not be more different. As James himself makes clear, he and Tristan's orientation to their new place is driven not by an attempt to achieve an inwardly focused and static familiarity with Barcelona, but by contrary and explicitly open and expansive impulses:

IS: Why did you choose to come to Barcelona?
James: Connection, inspiration, anonymity. For a whole lot of reasons
           . . . fashion . . . sensibility.

This account, then, is not an attempt to locate and describe the essential elements of place-making, but rather to examine one (perhaps atypical) case which draws attention to the many tensions at play in place-making practice. Neither, in this context, can Tristan and James' experiences be understood as in any way "typical" of contemporary dislocation. Forms of human mobility are as radically variegated in terms of the power differentials at play as modes of spatial experience themselves.[11] Indeed, their atypical, somewhat ambiguous status as something other than "tourists" or "migrants" or "locals," makes them particularly interesting subjects. On one hand, this is because the world is increasingly characterized by the movement of people and other connections between places, and on the other, because at the time when this research was conducted, they had lived in Barcelona for four years and this means that, in an everyday sense, they were also in the process of making the city into a kind of home.

The chapter proceeds, then, by examining how place is articulated in relation to mobile communications in the light of arguments that have been made in both popular and academic literature, on the impact of mobile communications on the concept and experience of place. Through this analysis a more nuanced account of place is opened up—one that challenges the supposed "disembedding" qualities of devices and dispels the notion of places as discrete and bounded. The chapter then goes on to provide a brief phenomenological account of James and Tristan's experience of Barcelona. A key focus here is the way in which they make Barcelona home by way of physical engagement with the place and with the people in it. Theirs' is not an abstract relation to place, but one in which they are fully immersed and which is articulated through everyday practical engagement and bodily movement about the city. Next, a challenge is mounted against the notion that technologies and media in particular are necessarily opposed to phenomenological processes of place-making and movement by way of an examination of how mobile communications are enrolled in such processes. Finally, an alternate understanding of phenomenological place-making practice is put forward that is able to account for the implication of technologies and the differences in scale that they produce. This alternate approach foregrounds the notion of familiarity that is so central to phenomenological geographers' understanding of relation to place. It is suggested that James and Tristan's experiences can be understood in terms of a rhythmic relation to the city composed of the familiar and the strange that incorporates both their relation to Barcelona and the technologies that they use to experience it. Without fully reconciling them, this focus on rhythm opens up the possibility of bringing a phenomenological understanding of experience into dialogue with contemporary accounts of sociality emanating from the studies of science and technology. In particular, with Actor-Network Theory (ANT), which famously includes technologies as actors in the constitution of the networks of people and, in its own terms, "non-humans," through which human subjectivity is produced.[12] Networks, as ANT scholars point out, are extremely messy, being composed of many heterogeneous parts, yet they also represent a kind of order that is achieved through repetition.[13] The combination of mess and repetition can also be understood in terms of rhythm.

MOBILE MEDIA AND PLACE

In Australia, mobile phone subscriptions doubled between 1997 and 2001, a period during which the meaning of mobile devices could be said to have shifted—from the specialized tool of select business, to a ubiquitous and everyday necessity for people from all walks of life. It was a rapid pattern of adoption that was mirrored across the developed world.[14] And this was a unique medium, not because mobile phones brought with them any radical new functionality,[15] but because they enabled a shift in the location in which an existing medium could be used. Yet this change in location and sudden ubiquity was perhaps as profound and radical as any development in function might have been, since it demonstrably reconfigured social relations both across space and in place. The advent of mobile communications drew attention to the interaction between place and media technologies in ways not witnessed since the advent of domestic media technologies like TV and radio. The technologies gave a novel (and somewhat literal) twist to Raymond Williams's concept of "mobile privatization."[16] If the growth of television and the supremacy of the private motor vehicle were the technologies that enabled "mobile privatization" at the level of the private family and the suburb for Williams, then the mobile phone further granularized the concept to the level of the private individual.

In response to this social and technological upheaval, popular commentators bemoaned the use of mobile phones in theatres, on public transport, in restaurants and cafés and in public space in general. Academic discourse took these concerns further, focusing on the often-tacit rules that govern interaction in particular places. The dramaturgical approach of sociologist Erving Goffman[17] became something of a touchstone for such studies as the "front stages" and "back stages" of social life appeared to be shifting fluidly in front of our eyes.[18] The regulatory "definition of the situation" composed of stages, props, costumes, actors and audiences was suddenly disrupted. Place had been "doubled."[19] Notwithstanding the often considerable nuance of such analyses, the attempt to illuminate the nature of the *complications* wrought by mobile communications technologies, arguably also brought with it a regressive *simplification* of what it is that constitutes places, particularly in the contemporary world. Rich Ling, for example, in his sociological account of "the cell phone's impact on society" provides a richly detailed account of the repertoire of bodily gestures that mobile phone users employ to mark their "removal from the social situation."[20]

To understand the mobile user as removed from a social situation, is of course, also to understand the social situation as itself being composed at a local level between visible and bodily present individuals. No doubt there may be, in particular instances, some attempt to keep phone-space separate from the local social situation, but it is certainly not always the case. Either way, the issue of removal from the local situation or otherwise, is likely to be more pronounced in the period during which mobile devices are first becoming ubiquitous. Following this, it is likely to recede, as devices themselves become part of any given social situation resulting in much more nuanced understandings of how phone space might intersect with local places. James's account, in the current case, recognizes both the rich variability of social situations and the relations between people, places and technologies of which such situations are (always) constituted.

For James, place is not simply one thing that might be "doubled" by bringing it into contact with another via mobile communications technologies. The complexity in the way in which he describes where he might use his phone puts the point eloquently:

James: Anywhere. I won't use it if I'm . . . if I'm in a queue at a bank. If I'm in a taxi, I just don't call from a taxi. I wouldn't use it . . . I wouldn't like be eating dinner and making calls, but if someone calls me and I'm eating dinner, I'll take it but I won't talk for hours. If I'm in a restaurant eating, I won't take the call.

IS: And if you're in a restaurant is that different from being in a café?

James: Yeah, depending on who I'm with. Like if I'm with my lover and we want to ring someone, then sure, I'll do that in a restaurant as well. But if I'm out with friends I won't make calls.

For James, it is not merely a case of mobile communications introducing a new element into a delineated, static and bounded place. James's description of the place in which he *might or might not make or take* a mobile call is already marked by radical contingency and fluidity that includes a heterogeneous array of both local and mobile mediated possibilities. Mobile communications devices are integral to the constitution of "the social situation." They are part of James's ordinary, everyday interactions. And on this level, phenomenological perspectives may prove particularly insightful since these provide a way of thinking about the experience of place which is capable of illuminating certain seemingly mundane bodily and habitual interactions of which mobile communications are now a part. To tease out how this is the case, it is worth first foregrounding the extent to which the phenomenological perspective on place accords with James and Tristan's experience of Barcelona.

PRACTICING PLACE

The picture of a road leading to a distant cottage seems easy to interpret; yet the road makes full sense only to someone who has walked on it. [21] (Yi-Fu Tuan)

One thing that stands out in Tristan and James's descriptions of their everyday lives in Barcelona, is the physicality of the place. Tristan explicitly foregrounds an intensely embodied relationship to the city in explaining why they chose to live in Barcelona:

And Europe . . . there's a lot of action here, it's a lot of fun. . . . And I really think that the beach culture was a bit of a huge pull for us. We're real swimmers and beach boys. There's not really any other city that's big enough in Europe that really has a beach and we need a city.

It also continues to be the case at least up until the time that they participated in this research. The layout of the city and the distinctive architecture of *Barri Gotic* and *El Raval*, where Tristan and James lived and worked, features heavily in descriptions of their spatial orientations to everyday life in the city, constructed in and through narrow twisting alleys opening suddenly onto unexpected squares, monuments, churches and hills. In part this may have to do with the distinctive architecture of the city. But there are also two other likely reasons why both Tristan and James's descriptions of their life in Barcelona are so infused with the physicality of the place. Physical dimensions, landmarks, geographic icons and the sensual qualities are particularly apparent during early experiences of particular places. The novelty of their chosen city of residence for James and Tristan is useful since it enables us to see what might otherwise remain hidden. In phenomenological accounts, the physicality of environments remains all-important in structuring the everyday experience of place, but knowledge of these is not characterized by cognition, which might enable it to be articulated as such. Rather, as places become familiar and integrated into everyday routines, a bodily knowledge of place emerges "which is reducible to a sort of co-existence with that place."[22] "Cognition," according to David Seamon, "has a role in everyday movement, particularly if that movement is new, novel, or occurring in an unfamiliar environment. A larger portion of movement, however, arises from the prereflective sensibility of body-subject."[23] As the body learns its environment, the physicality of place slips below the threshold of accountability.

The second important aspect of James and Tristan's physical relation to place here has precisely to with the establishment of "everyday movement." In this respect, their approaches to the city are quite different from those of the tourist. Their daily round of activities does not consist of a schedule of landmarks and sights to see, but is rather oriented towards the formation of an everyday life in a new place. There are a series of obvious ways in which this is the case. They must grapple with new languages, and navigate a new social, spatial and temporal milieu. And while these are in part carried through via what might be termed abstract, cognitive approaches such as maps and books, a much larger proportion is carried out via direct bodily engagement with the place. The establishment of their business and social life in Barcelona means speaking the local languages in everyday situations, navigating the roads, alleyways, buildings and squares of the city, and tuning to the temporal rhythms of the city.

The physicality of their interactions with the city is structured around the establishment and running of a business. Tristan and James' business is a specialty sweet shop in *Barri Gotic*, five minutes walk from their apartment in *El Raval*. Like other businesses in Barcelona, the shop is open from 10 a.m. until 2 p.m. and then again from 5 p.m. until 10 p.m. They do alternate days, each working, on average, approximately eight shifts a week, boiling, stretching and cutting hot brightly colored caramel in a smallish kitchen at the back of the shop. The work is physical, and boiling sugar keeps inside temperatures high even on a cool day. There is also considerable time pressure associated with the work, as there is a limited window in which this shaping can be carried out before the candy becomes too cool and hard to be pliable. They generally have one or two local staff working on the shop floor while one or both of them work out the back. Their time off is spent swimming at the beach in summer and at the pool in winter, shopping, mostly for food or household items, meeting friends and visiting art galleries. Because they work alternate shifts at the shop much of this time is spent apart from each other, walking or riding bicycles through the cobbled streets of the city. Tristan and James establish Barcelona as a place for themselves, as a home (albeit a temporary one) by moving about it, and establishing bodily routines that explicitly connect them to the city. They are developing what Seamon calls "bodily knowledge" or Tuan terms "spatial ability"—an everyday pre-reflective relationship to place.

DISRUPTIVE MOBILITY

Movement about the city is central to the making of James and Tristan's everyday life in Barcelona and is to this extent in accord with phenomenological-geographic conceptions of place-making. But not all movement is equal in phenomenological accounts, which often effectively valorize particular modes of movement over others. The movement of the body, its movement around local neighborhoods, is seen as impacting positively on our experience of place, while the "upheavals" of car and train journeys, and particularly the dislocating effects of jet air travel, have quite the opposite effect.

In establishing themselves in Barcelona, however, it is immediately obvious that it is not just a question of local unmediated movement for Tristan and James. The physicality of their descriptions of movement about the city extends also to various technologies that enable this movement. The cobbled, often narrow and densely peopled streets were, for example, one of the reasons, according to James, that "no-one has a car here" and "everyone rides a bike." Mobile communications technologies were prominent in this respect also. They do not, of course, enable movement in the sense that a bike or a car might, but in many ways their role in structuring both James and Tristan's movement about Barcelona is even more critical. Mobile communications are central to the establishment of the business, which is the express reason that James acquired a mobile phone shortly after arriving in the city since he "needed to set up the business and to call real estate agents, to make connections with clients."

Mobile communications devices are, in this sense, immersed from the beginning in the regimes of everyday local interaction through which James and Tristan create their life in Barcelona.

The role of mobile communications in structuring Tristan and James's movement and orientations to the city remain apparent in a number of important ways. First, mobile communications devices are central in locating themselves in, and as they move about the city, in a very literal sense. They were used to send and receive directions to particular places in an unfamiliar setting. Second, as both business and life partners who work alternate shifts in the shop, much of their everyday lives are spent apart from each other and much of their mobile phone use is accordingly conducted between the two of them. Mobile communications were frequently observed to simply locate Tristan and James in relation to one another. The issue of locating oneself in relation to a mobile communications interlocutor is a commonly observable phenomenon, which Eric Laurier has explored in some detail.[24] Whereas Laurier locates this positioning as a function of the strategic spatial circumstances of mobile phone conversations, James and Tristan's interactions were not of a straightforwardly functional nature. Frequent contacts between Tristan and James testify to a much more sensual relation to place in which strategy and function play little or no part. The multisensory nature of contemporary devices enables a repertoire of phatic interpersonal communications about place, exemplified in describing the sending of photos via these devices. According to Tristan (emphasis added):

> I mean this morning I took a photo while I was having breakfast and sent it to James and said, "It's a shame you're not *here* having coffee and toast," you know?

And similarly James (emphasis added again):

> I send one occasionally, like something funny of someone doing something stupid *on the street* or *some beautiful place*.

In each case, a sense of location is shared via the device, to convey something of the sensual qualities of that place with the other.

Third, and particularly importantly, mobiles were also about developing a sense of place that was manifestly social. On first arriving in the city, James and Tristan knew just one person in the city between them and spoke neither the local Catalan nor Spanish. Since then they have become fluent in Spanish and developed a wide network of friends. Tristan says that his mobile interactions are "about 90 something percent social." And importantly, these social interactions were about coordinating the physical movement of people about the city. When Tristan is asked what he most uses mobile communications for, his answer is clear:

Tristan: When there's quite a few people, as a way to kind of, to corral everyone together. As in, "where is everyone now? We're meeting at this corner."
IS: So to organize people?
Tristan: Yeah, to organize for people to be in a certain place at a certain time.

Mobile communications technologies are here closely integrated into place-making practice and in the same phenomenological sense that the connections they make through the city are. That is, mobile technologies are entwined with practical engagement with the city on an everyday social and business related level. Nevertheless, James and Tristan's experience of the city here also begins to depart from phenomenological geographers' accounts of place-making, which have often tended to privilege a simple unmediated connection between individuals and communities such that "places are fusions of human and natural order and are the significant centers of our *immediate* experiences of the world."[25] Media of any sort are, in this sense, conceptualized as the very antithesis of authentic place-making practice. Building on Relph's conceptualization of "place and placelessness," David Seamon makes this point explicit:

> Today, in an era of mobility and mass communications, people easily transcend physical space and readily compare and switch places. . . . Many people are footloose and feel no attachment to place. At the same time, technology and mass culture destroy the uniqueness of places and promote global homogenization.[26]

Seamon here puts the disembedding effect of technology and mobility effectively down to two key points. One is explicitly the intervention of technology itself. The other is implicit, and is a question of scale. While movement at the local embodied level is valorized as having an embedding effect, movement across larger spaces—the "mobility and mass communications" referred to above—have quite the opposite effect. But there is another possible axis along which questions of embedding, at-homeness and other relations to place can be considered, that is neither primarily a question of technology nor of scale. That is, these can be considered in terms of the rhythmic processes of everyday life and the balancing of the familiar and the strange.

It is worth at this point returning to Maurice Merleau-Ponty's *Phenomenology of Perception,*[27] since it provides the philosophical basis for Seamon's conceptualization of place and gives careful consideration to the ways in which technologies associated with movement may be enrolled in spatial knowledge and motility. In a well-known passage from that work, Merleau-Ponty writes of the way that, as a driver, he "knows" the breadth of his car and how a blind man knows the extent of his "stick." He refers here, of course, not to an abstract cognitive knowledge, but of the practical, pre-reflective bodily knowledge built up through habitual use: the same mode of knowledge to which Seamon refers in articulating the experience of place. To use implements in this way is to come to in-habit them, or to take them into the body:

> To get used to the hat, a car or a stick is to be transplanted into them, or conversely, to incorporate them into the bulk of our own body.[28]

The phenomenological focus would seem, then, to offer a useful way forward in considering technologies that are intimately connected with bodies, their movement, perception and interactions. For both Tristan and James this mediation is clearly articulated through their use of mobile communications technologies. They respond, by way of practical engagement, to complex spatial co-ordination concerns that are ultimately locally and physically instantiated—"at this corner"—*through* their mobile devices. Thus, as is the case with Merleau-Ponty's car or blind man's stick, the extent to which James and Tristan are "used to" their devices becomes a relevant factor in understanding the phenomenological aspects of their place-making practice. And like their relationship to Barcleona itself, their attitudes to mobile devices are also about constructing a rhythm out of the tension between the familiar and known and the sense of "connection and inspiration" associated with the newness of the place.

CONTINUITY AND CHANGE

One of the first things to note in discussing this tension is that Tristan and James's move is not all about Barcelona's novelty. Asked why they had chosen to come to Barcelona, Tristan specifically highlights continuities between the city and their life back in Australia:

> I don't know, it feels to me like Spain's probably got the closest culture to Australia, what I'm used to, like the way of living, eating well, being outside, having a nice time.

While they seek, according to Tristan, "an opportunity to live in another country and learn another language and experience another culture," there is also a sense in which the connections between this newness and the familiarity of home are highly relevant. Not least they are so for this discussion because such connections and continuities are now *by necessity* constructed in and through various forms of media: travel books before making the move to Barcelona and various communication channels—including mobile media—to "keep in touch" afterwards. From the beginning, then, there is a tension here between continuity and change that is played out in quite specific ways. Their descriptions of Barcelona are cut through with the language of experience and immersion common to the phenomenological experience of place. They do not want to simply *see* Barcelona, but to "live in," "experience," and "learn" the "ways of living" that the city has to offer them. Mobile communications are explicitly enrolled in these ways of living. More specifically, James and Tristan's respective relationship to mobile communications devices can be viewed in terms of a similar set of rhythmic exchanges between the familiar and the novel as their developing relationship with the city itself. At the time during which fieldwork was carried out, Tristan and James had been living in Barcelona for four years, having arrived in 2002. Their attitudes to mobile communications were, at the time of their arrival, quite different. James had a long-standing close relationship with them, saying:

> I used to have a mobile phone bill of five or six hundred dollars a month in Australia for business, I was running a business, it's busy . . . I love them.

James got a mobile almost immediately for the purposes of setting up the business—establishing a routine in relation to the rhythms of the city itself. Tristan, on the other hand, was, by his own reckoning, "never really that interested in them." Before moving to Barcelona, he had never owned a mobile phone, and did not acquire one until they had been living in the city for three years. Tristan and James's respective attitudes to the device have been to some extent imported from Melbourne to Barcelona, as have their orientations to the new city itself. Their attitude to subsequent phones is also dependent on their previous interaction with devices. Thus, James indexes his love of the phone to a long history with devices in general: "I like my cell phone, always have;" and with specific devices in particular: "I'm a big Nokia fan. I've used Nokia for years." James is from the beginning "used to" and familiar with mobile communications devices and they slip easily into his experience of Barcelona. Tristan's experience in this respect is also interesting, since he is very much in the process of *getting used* to his phone:

> I never felt like my life was lacking when I didn't have one and now I have one I find it hard to imagine what I was doing when I didn't have a phone.

As James's device was from the beginning, so Tristan's mobile has quickly become integrated into the rhythms of his everyday life in the city. The default for his mobile is always "on"—switched on and on his person. Tristan's description of those times when he does not have his phone is in this respect enlightening, because he describes a situation not only where he does not have his phone with him, but one in which he has almost nothing:

If for example I'm going to the beach, I just don't take it. I'll just take some money in my pocket and that's it. So I just don't take anything with me.

In this formulation, Tristan's mobile device is second only to the universal exchange commodity—money. The interaction between bodies and devices achieve an ordinary, predictable status not by "reflecting" but by becoming embedded in habitual temporal and spatial routines because, like money, mobile communications devices are enrolled in constituting them.

CONCLUSION—SOCIO-TECHNICAL RHYTHMS

Despite being conducted in the late 1970s, well before the widespread diffusion of mobile media, Seamon's work resonates with the temporal and spatial rhythmicity observed in relation to James and Tristan's mobile interactions. Seamon writes of what he calls "place ballet," built up through everyday time-space routines, "a set of gestures and movements which sustain a particular task or aim."[29] It is a conception which connects explicitly with the specter of bodies moving through public space as if attuned to complex rhythms that James Katz evokes in his examination of the "choreography of mobile communication."[30] The choreographic metaphor articulates the embodied, rhythmic nature of everyday spatial experience and is suggestive of the means by which technologically mediated interaction connects with and tempers the movement of bodies in local, immediate situations. Seamon's pre-conscious body subject is, after all, governed at a habitual, *automatic* and *mechanical* level. Tristan and James's everyday movements are set to diurnal, biological, social and mechanical rhythms, expressed through a practical everyday engagement with the city, which is in part articulated by the technological rhythms of their mobile devices. The mobile communications system is enfolded into the layers of predictability and contingency which characterize inhabitation of the world by an embodied subject.

In their rhythmicity, both Tristan and James's experiences are congruent with Seamon's "place ballets." Certain aspects of the experience of place are entirely embodied and become a part of the backing track sustaining everyday life. This chapter has demonstrated how, through ubiquity and entry into the domain of the everyday, mobile technologies may become enfolded into pre-conscious regimes of interaction becoming part and parcel of a bodily perceptual apparatus through which place is experienced. Indeed, despite an apparent antipathy toward media technologies, even in Seamon's study, the distinction between mediated and otherwise is difficult to sustain, such that for one of his subjects, the automatic dialing of a telephone number takes its place alongside other habitual orientations to space like "going into a room" and "reaching for a towel."[31] In this study, unfolding in what is certainly a more technologized world than that of thirty years ago, the imbrication of technologies in mediating the spatial arrangements of the everyday as a crucial constituting factor in the composition of spatial experience has only become more evident. Moreover, as these technologies are enrolled in the choreographing and performance of the everyday, it is precisely they that take on the sort of submerged significance that Seamon and other phenomenological geographers might ascribe to an unmediated "existential insideness," or "at-homeness" of attachment to place. Such attachments are indexed not to the extent that experience of place is technically mediated or otherwise, nor to the scale at which movement unfolds, but to the rhythmicity (replete with repetition and deviation) of movement and encounter.

NOTES

1. David Harvey, "From Space to Place and Back Again: Reflections on the Condition of Postmodernity", in *Mapping the Futures: Local Cultures, Global Change*, eds. J. Bird et al (London: Routledge, 1993), 2–29, presents perhaps the best-known and influential argument along these lines.
2. Yi-Fu Tuan, *Space and Place: The Perspective of Experience* (Minneapolis, MN: University of Minnesota Press, 1977).
3. Edward Relph, *Place and Placelessness* (London: Pion, 1976).
4. David Seamon, *A Geography of the Lifeworld: Movement, Rest, Encounter* (London: Croom Helm, 1979).
5. Seamon, *Geography of the Lifeworld*, 41.
6. The transformation of space into place through everyday interaction is a common tenet of phenomenological geographic perspectives articulated most clearly in Yi-Fu Tuan's *Space and Place*.
7. In mobile communications literature this conceptualization has appeared perhaps most explicitly in Leysia Palen, Marilyn Salzman and Ed Youngs, "Going Wireless: Behavior and Practice of New Mobile Phone Users," *Proceedings: CSCW* (New York: ACM, 2000), 201–210.
8. This perspective is often most apparent in phenomenological geographic accounts of place deriving from Heidegger, especially chapter IV on "Building Dwelling Thinking" in Martin Heidegger, *Poetry, Language, Thought*, trans. A. Hofstadter (Harper Colophon: New York, 1971).
9. Relph, *Place and Placelessness*.
10. Seamon, *Geography of the Lifeworld*.
11. See David Morley, *Home Territories: Media, Mobility and Identity* (London: Routledge, 2000); and John Urry, *Sociology Beyond Societies: Mobilities for the Twenty-first Century* (London: Routledge, 2000).
12. A clear exposition of the actor-network approach is to be found in Bruno Latour, *Reassembling the Social: An Introduction to Actor-Network-Theory* (Oxford: Oxford University Press, 2005).
13. Steven D. Brown and Rose Capdevila, "Perpetuum Mobile: Substance, Force and the Sociology of Translation," in *Actor Network Theory and After*, eds. J. Law and J. Hassard (Oxford: Blackwell, 1999), 26–50.
14. International Telecommunications Union data, http://www.itu.int/ITU-D/ict/statistics/; accessed 10 February, 2011.

15. See, for example, Chantalle de Gournay, "Pretense of Intimacy in France," in *Perpetual Contact: Mobile Communication, Private Talk, Public Performance*, eds. J. Katz and M. Aakhus (Cambridge: Cambridge University Press, 2002), 193–205.

16. Raymond Williams, *Television: Technology and Cultural Form* (London: Fontana, 1974).

17. Particularly, see Erving Goffman, *The Presentation of Self in Everyday Life* (New York: Doubleday Anchor Books, 1959); and *Relations in Public: Microstudies of the Public Order* (New York: Harper, 1971).

18. Mobile communications studies taking up Goffman's ideas include Rich Ling, *The Mobile Connection: The Cell Phone's Impact on Society* (San Francisco: Morgan Kaufman, 2004); and Ged Murtagh, "'Seeing the Rules': Preliminary Observations of Action, Interaction and Mobile Phone Use," in eds. B. Brown, N. Green and R. Harper, *Wireless World: Social and Interactional Aspects of the Mobile Age* (London and New York: Springer, 2002), among many others.

19. Emanuel Schegloff, "Beginnings in the Telephone," in *Perpetual Contact: Mobile Communication, Private Talk, Public Performance*, eds. J. Katz and M. Aakhus (Cambridge: Cambridge University Press, 2002), 284–300.

20. Rich Ling, *The Mobile Connection*, 134.

21. Tuan, *Place and Space*, 22.

22. Maurice Merleau-Ponty, *Phenomenology of Perception*, trans. C. Smith (London: Routledge & Kegan-Paul, 1962).

23. Seamon, *Geography of the Lifeworld*, 89.

24. Eric Laurier, "Why Do People Say Where They Are During Mobile Phone Calls?", *Environment and Planning D: Society and Space* 19, no. 4 (2001), 485–504.

25. Relph, *Place and Placelessness*, 141, emphasis added.

26. Seamon, *Geography of the Lifeworld*, 91.

27. Merleau-Ponty, *Phenomenology of Perception*.

28. Merleau-Ponty, *Phenomenology of Perception*, 165.

29. Seamon, *Geography of the Lifeworld*, 54.

30. James Katz, *Magic in the Air: Mobile Communication and the Transformation of Social Life* (New Brunswick and London: Transaction, 2006), 39.

31. David Seamon, "Body-Subject, Time-Space Routines, and Place Ballets," in *The Human Experience of Space and Place*, eds. A. Buttimer and D. Seamon (London: Croom Helm, 1980), 153.

Part IV

Bodies, Screens, and Relations of Place

10 Going Wireless
Disengaging the Ethical Life

Edward S. Casey

I.

While it may be foolhardy to presume to grasp the full consequences of new technologies, there are questions we can ask that will elucidate a certain taken-for-grantedness about life—the very ways of life that may undergo the greatest changes.

To begin with: What does wireless technology mean for us as embodied human beings? At first glance, one would think that such technology offers a highly desireable state: aren't wires, plugs, and other associated paraphernalia excessive and unnaturally binding? Don't they encumber the spontaneous movements of the living body, which would like to make its own way unfettered by accessories and attachments that, quite literally, tether it to a certain place?

In fact, there are fundamental features of embodied existence that suffer neglect in a wireless world, and change the experience of ourselves, each other, and the living world. First among these is the irreplaceable value of being with other humans in a face-to-face manner, thanks to our being in the same place. There simply isn't any substitute for the nuanced reading of the other's face and, indeed, her or his whole body. The expressions and dialogues made possible by being in the presence of a person are of an intricacy and scope that simply cannot be experienced otherwise.

You may ask: but what about encounters in which both parties can view the other in real time with perfect color and in three dimensions via digital imaging? Certainly, we can see the other's face and maybe his or her body as well, and these visuals can facilitate dialogue. Hence the increasing use of such technology in doctoral examinations where one or more of the official readers is unable to be present at the actual place of interrogation. Nevertheless, there are inherent limits in this situation—limits that restrict the range of "dialogical interaction."

So long as there is an overriding task at stake, things go well enough: at the level of the topic (question, point, content) a certain conversation can occur. But we miss the subtle cues that come with the corporeal presence of the other to me and myself to him that aren't entirely visual. When we are able to read one another's breathing and speech patterns, skin tones, nervous energy, placidity, and so forth, we may pick up (and mutually amplify) subtle signals of uncertainty, sparks of interest, or resistances that may open both of us to an entirely new (and perhaps vital) direction in our exchange.

Issues of a different magnitude also arise. For instance, the illusion that another person is available to me merely because I carry a cellphone or a laptop. I tend to assume that everyone in my address book is "with me" everywhere I go. Instead of setting up actual meetings or live conversations, I rely on the assumption that I can always "check out" what the other is up to anytime I want and vice versa (hence, I consult my voice mail and messages fairly constantly). The more people I know with the same technologies, the more we both experience being "with each other" no matter where either of us actually is, and without the kind of anticipation and preparation that comes with looking forward to a meeting. A reflexivity sets up and reinforces the illusion that I am as connected with my contacts, and they with me, as we would have been had we met in person.

Certainly, there is no question that such accessibility to others is *useful* on many occasions, and perhaps even necessary on certain others, as in the case of an emergency. And it is true that I cannot be everywhere at once. But the technologically mediated presence of another person brings with it the presumptive illusion that I can, without cost or loss, replace the concrete other with his or her icon.

Who is it, or perhaps better said, *what* is it that I reach in this way? Not the full person but his or her virtual image; not the flesh but the figure; not the person who is changing in thought or feeling *because of our very interaction* but his or her imago or wraith. The fact that it takes some very particular effort to achieve this direct intersubjectivity—that either she or I, or both of us, must move our bodies into the other's actual ambience—is not merely a matter of inconvenience. It signifies the indefeasible value of being *with* the other for certain purposes: not only for the sake of experiencing myself in the presence of others but for courting a potential lover and teaching students, going to a speech by a politician and being part of a political demonstration.

These events are worth displacing our bodies to get to—even going to extreme lengths to do so, as when people travel half-way across the continent to hear Archbishop Tutu address a peace rally in New York. The necessary displacement of my body on such occasions signifies the irreplaceability of that same body. To call in or type in my support for the rally as if this were tantamount to being there betrays a confusion between the operation of a virtual self and the presence of a corporeal self. To come to the right place requires the actual movement of my lived body *to that place*. One singularity rejoins another.

II.

Indeed, place is not an incidental dimension. Beyond its importance as the material condition of being face-to-face with the other person and its indispensability to being a full part of the public realm, place is at stake in another advance made possible by a wireless world: the locational power of global positioning systems (GPSs). With GPS capabilities, information and utility vie with presence—in this case, *presence in place*. It may be helpful to know just where I am located in cartographic space as I drive through a region with which I am not familiar: I can tell how far I am from a certain definite destination, as well as how to reach it most efficiently. Certainly, the use of GPS to locate lost children or animals is of undeniable value.

But if I go into wilderness with a GPS backup to ensure my rescue if anything life-threatening should occur, have I not surrendered one of the most valuable aspects of going to the wilderness in the first place? Am I subtly less present, less alert to situations of danger and thus less attuned to the intelligence those situations would solicit? Will I become less able to deal with risks elsewhere in life? When I know I can call upon a service such as OnStar for roadside service in the event of a flat tire, will I become less competent at changing a tire? Will I bypass a stranded motorist or be reduced to making a call for him too, joining him in mutual helplessness? Do such new dependencies induce a creeping infantilism, a more pervasive lack of bodily "know-how" that feeds into and reinforces a world increasingly defined by anxieties, fears, and insecurities?

A more complex case is the use of GPS to determine the exact position of illegal "aliens" who have crossed the US-Mexico border. Yet it is not at all clear that such migrants are dangerous to American society, much less that they should be considered objects of prey. And it is a slippery slope for authorities mandated to keep watch on potential "evil doers," when surveillance takes an easy leap to around-the-clock tracking by satellite.

Beyond such benign as well as dubious uses, the basic fact is that GPS does not tell me anything significant about *where I am*. That is to say, *in what place I am standing*, or where or in what concrete place a given child or migrant may be. My GPS device may well indicate that I am at 45 degrees latitude and 32 degrees longitude, or four miles east of Longmont, but what does such information tell me about the very place in which I find myself? What does it tell me about *this* very stretch of desert around me, *those* mountains I see from that same desert place, the way *this* road on which I am driving *feels* to me as I go over it in my aging Honda Civic? Little, if anything. I have found my site in space but lost my way in place.

A gain in one respect brings with it, ineluctably, a loss in another. But this is not a simple "trade-off" that can be easily judged or justified as such. Nothing goes for free in the wireless world, and not just because I almost always have to pay for the pertinent technology. More importantly, I lose touch with what is essential to and a counterpart, of my lived body: place. For *there is no body but in place*. I am nothing if not this body—it is the primary requirement of my existing at all. And I am also nothing if not in place. When I use wireless technologies that take me out of place, no matter how convenient or practically valuable that move may be, I am moving into a disembodied experience that deprives me of the very basis of my personal and interpersonal identity.

Further, if the ground of my identity is undermined by my use of wireless technologies, then the autonomy at stake in such identity is also threatened. I may claim a certain freedom of choice in employing such technology: I can call my friend Maria at (almost) any time I like, day or night, no matter where I am, but this freedom is not equivalent to personal autonomy. For one thing, such a freedom is circumscribed by certain definite conditions of use, for example, that the transmitting towers near me are sufficiently strong to support my employment of a cell phone. For another, even when functioning successfully, such freedom falls short of autonomous action in the basic Kantian sense.

Indeed, we are compelled to ask whether such limited freedom of choice can *ever engage the ethical life* in any significant regard. Instead of actualizing autonomous self-shaping action, it operates at the level of what Kant calls the "hypothetical imperative." My attempts at relating from a distance are necessarily predicated on an if-then logic: if I want to accomplish X, then I must employ Y. Thus, if I am out walking on the beach or sitting in a café in Santa Barbara, and have an inspired thought that prompts me to consult with my friend Dick in Phoenix, then I must call him on my cell phone or e-mail him from a laptop in range of WiFi. Left untouched by my assumptions are such vexing questions as these: Should I contact him right now by means that are convenient for me but possibly intrusive and upsetting to him? Am I respecting him as a person when I reach him by such highly mediated instruments as cell phones and computers? Might he feel diminished by my assumption that he is always available to me by these instruments? This assumption amounts to a reduction in his autonomy, which is as fragile as it is deep, as valuable as it is cheap.

Furthermore, when we circumscribe the bulk of our daily existence with "information management," or communications-at-a-distance, we lose the very fabric of responding to a world that exists around us and that calls us to do some very specific things. Perhaps most importantly, we automatically remove ourselves from directly experiencing the consequences of our words and deeds. Our sense of self, if exercised wholly as a function of what we want, when we want it, risks becoming perilously arrested in what may be developmentally appropriate for a toddler, but not for an older child who must learn difficult lessons from her or his actions in an actual social situation, with animals, or in a wilderness area; or, for a mature adult who must respond to, care for, and guide others.

III.

A third area of questioning concerns *time*. Wireless technologies are more efficient means of doing what previous technology had aspired to do from the very beginning—namely, to bring all of time into the present moment of reception/perception. In the case of video technology, a cascade of images and words can be made to appear *on my screen right now.* I may receive multiple views made available to me by wireless home security cameras, photos sent via e-mail or camera phones, or witness live webcam casts from around the world. In effect, the past and future horizons of time are disregarded in radically new ways.

In Merleau-Ponty's phrase, a "freezing of being" has occurred. This is not to say that these horizons have been cancelled or eliminated. They continue to exist for *any* event, whether in attending to the modest screen of my personal computer or in scanning the closely coordinated screens of the GPS in the Border Patrol Building at Nogales, Arizona. But past and future are suspended in the moment of the event of viewing. True, they can be remembered or, in extremity, posited as having had to be there; but their organic link to the present is severed. I see what is happening just now, but I am not given access to how this *now* arises from its own idiosyncratic "ground" in the immediate or distant past, nor may I gain access to how it leads into the unfolding future. In short, the kind of *becoming* that is intrinsic to the full experience of time is short-circuited; it is reduced to what-has-now-become—and only insofar as it is set before me on a liquid crystal display, plasma screen, or other representational surface.

While many of these observations hold true for wired as well as wireless technologies, the latter represents something at once quite special and yet quite familiar since the advent of early modernity. Wireless devices of whatever sort—Internet, handheld phone, camera, GPS, electronic devices for locating lost children and pets, instant messaging, e-mail, cars with OnStar GPS, etc.—share the basic trait of *action at a distance.* They instantiate information (whether in images, words, or other symbols) from a nearby or distant, and often unknown, site in space so seamlessly and efficaciously that we are only dimly aware, at best, of the actual transaction or transmission. Even time itself can be distanced, as when a filmed event is televised at a later moment in time.

Such non-contiguous spatio-temporal effectuation, for all its formidable impact on our lives, can be seen as a man-made version of what seventeenth-century physics conceptualized as the instantaneous effects of such invisible but potent universal forces as gravity or light. As these latter act across space and time with definitive causal power, so too does wireless technology transmit information with equally effortless aplomb.

The difference is that the action at a distance made possible by wireless devices concerns spatial and temporal effects not of universal forces but of linguistic or imagistic symbolic systems. Such information is only conveyed to human beings, indeed only to those humans who are familiar with the representational system in which such information is coded. The causality of this system is almost solely effective in cultural contexts that maintain their own historicity and ideology. The implications of such privileging and narrowing are obvious: What will become of the need to know people unlike yourself, to feel empathy for others and respect for differences, to know and care for animals, plants, and places on earth?

IV.

As wireless technologies wend their way pervasively and subtly into our contemporary lives, they deeply affect the human experience. This happens most forcefully in the three primary areas of personal autonomy, place, and time. Moreover, these areas are internally connected. To lack place in the manner I have described is closely connected with the shrinkage of time to the current moment. In both cases, the lived body is ignored, or at least significantly restricted, in its potency.

Using a wireless device, it is as if I am moving with *part* of my body—or with an *objectified* body—which is not only less than the lived body but a body of a very different kind. Further, not knowing my way in place and thus being confined to a stringently defined "site," I am not open to the full embrace of time. The tie between a given site and what is happening *now* is very close, since a site appears in the present moment alone. Both become radically reduced, pinpointed versions of space and time. "Sitedness" and the "present tense" are distinguishable but *not separable* epiphenomena of the wireless age, and act to undermine the freedom of people living in this same age.

For a person to be genuinely autonomous, he or she must be capable of comparatively uninhibited action in place and time. To be the source of your own freedom, to write your own ethical laws (albeit of universal import) calls for movements in a full-fledged place—no shrunken site will do!—as well as for the perdurance of continuous time to allow one's own *becoming*. None of this is possible in ungrounded and isolated instants and sites of sheer representation. Hence, any claim to autonomy is baseless and one's actions remain ineffectual. If any free actions occur, they consist in the motions of thought alone—in the freedom to think X or Y—and not those of one's lived body: a body that requires a place in which to emerge and a time in which to develop.

Only such bodies in genuine places and in doubly horizoned times (with a past and future) can realize the dialogical expressivity to which I referred earlier in this chapter. Without sufficiently generous parameters of place and time alike, we are confined to communication construed as the transmission of items of information across an indifferent space and an apathetic time. There can be no doubt that wireless technologies not only facilitate such transmissions, but also, with ubiquity, allow them to become the norm. Thus, the questions we must ask are these: Are we willing to consider the general fate and specific price of such disembodied and displaced efficiency? Even more importantly, are we willing to learn the ethical value of living in our bodies, in places, and in time? Are we willing to engage ourselves in the very conditions that, above all else, make us human?

11 Parerga of the Third Screen
Mobile Media, Place, and Presence

Ingrid Richardson and Rowan Wilken

"Where does a parergon begin and end."[1]

This chapter attempts an intervention into debates about place and mobility which, as Tim Cresswell notes, tend to be "marked by disagreements between those who see mobility and process as antagonistic to place and those who think of place as created by both internal and external mobilities and processes."[2] Increasingly central to, but often overlooked within, these debates are body-technology relations, especially as they involve mobile media devices. In this chapter we consider the relation between place, the built environment, embodiment and modalities of presence in the context of mobile screen media. Drawing on the work of Edward S. Casey, we explore the relation between the body and mobile screen interface, and the way that the two work in tandem to alter our everyday experiences of place and presence.

To begin, we suggest that a critical interpretation of our perceptual and bodily engagement with screens can be usefully supplemented by tracing the various tropes and "body-metaphors" that are deeply embedded in our experience of screen interfaces. In particular, this focus on the body-screen relation reveals an affinity between windows, frames and screens, their place-making effects, and the complex ways we "turn" to them with varying degrees of attention and distraction.

The significance of the window, the frame, and more recently the screen, as markers and determinants of place—and of the "here-there relation" in Casey's terms more generally—cannot be understated. As Anne Friedberg notes, the contemporary screen has been instrumental in modifying our experience of place and (tele)presence.[3]

Yet, we would argue that particular attention needs to be paid to the specificities of our engagements with small, mobile screens (as distinct from, say, large, fixed screens). Today, our collective habits and routines include portable and handheld screens that can be carried with us, in our hands, pockets or bags, effectively mobilizing an intimate body-screen relation in a variety of different contexts. In such situations the conventional screen-window relation, and the place-making effects of the screen as "window-on-the-world" are fundamentally unhinged and destabilized. This chapter seeks to examine critically the screen-window-frame relation in the contemporary context of mobile screen use, and it suggests how our engagement with the mobile media screen is changing our experience of place, presence, pedestrian space and the built environment.

POST-PHENOMENOLOGY AND BODY-TECHNOLOGY RELATIONS

In developing this examination of body-technology-place relations, we are adopting what can be described as a "post-phenomenological" position. That is, while phenomenologist Merleau-Ponty's approach tends towards the notion of a typical, common or—by some accounts—essential mode of embodiment with a more or less universal "fit," Don Ihde develops a post-phenomenology which supposes instead that body-technology relations translate differently across cultures and their specific ways of perceiving, knowing, and making the world. Thus, in this chapter, our examination is anchored in a theoretical perspective, that we have articulated elsewhere, which acknowledges the ways by which the dynamic shaping of our corporeal schemas is under continuous modification by artifacts, tools, techniques and more complex technological ensembles, which are always-already embedded in a palimpsest of cultural milieus and collective habits.[4] In its post-phenomenological focus, then, our approach is framed by the premise that every human-technology relation is a body-technology relation, invoking certain kinds of being-in-the-world, and ways of knowing, making, and dwelling or inhabiting that world. Merleau-Ponty famously claimed that the body "applies itself to space like a hand to an instrument,"[5] an "application" that depends as much on the specificities of perception and bodily movement as it does on the materiality of the tool-in-use. The term corporeal *schema* refers to a being-in-the-world not determined by the boundaries of the material body; rather, our corporeality extends and withdraws—changes its very reach and shape—in its dynamic apprehension or *handling* of tools and things in the world. Merleau-Ponty argued that this schematic is inherently open, allowing us to incorporate technologies and equipment into our own perceptual and corporeal organization. In other words, our actions are not determined by intention alone, but also by the material environment; tools and things co-constitute reality and thus inevitably *co-shape* perception and experience of place.

For example, learning to drive a car involves learning to function according to the spatial organization and limits of the vehicle, its speed, the corporeal network/schematic of the hand-wheel-direction vector and foot-brake-deceleration continuum, and so on. Within the material shape and capacities of the car, we adjust our physical deportment, spatial orientation, and our entire physical relationship with the world. Initially, learning to drive involves constant attention and concentration, because we must consciously orient our bodies towards the unfamiliar spatial and motile logic of the car, but after some practice driving becomes habitual, and our being properly "dilated." In this way, the car becomes an aspect of our embodiment, part of our repertoire of proprioceptive skills; we have appropriated the body of the car as a temporary body or quasi-body, a supplemental being-in-the-world.

Building on this, Ihde's post-phenomenological approach focuses on the body-technology relations that emerge from particular cultural milieus and collective habits, and the way in which various technologies "transfer" asymmetrically across cultures. Ihde develops the generative concept of the relation from Heidegger's earlier claim that the relation *per se* is ontologically primary. That is, for Heidegger, the term relational ontology means that material objects and human being-in-the-world are not separate entities which are then *a posteriori* gathered up in a relation; instead, the relation itself is *a priori*, such that our separation of human and technology, thing and world, subject and object—or indeed this applies across the whole range of oppositional couplings—is heuristic and after-the-fact.[6] With this understanding of relationality in mind, Ihde differentiates between various human-technology-world relations, each of which indicates a particular mode of *being* in relation to equipment. These relations are based on the correlational schema Human <=> World: the upper line of the bi-directional arrow indicates the "first intentionality" or the directedness of experience towards the world, while the lower "'reflective' intentionality is the movement from that which is experienced towards the *position* from which the experience is had [. . . precluding] any simple talk about 'objects themselves' or, equally, 'subjects themselves,' since whatever falls out in the analysis is about *relations* between experiencers and experienced."[7] Of particular relevance to our corporeal experience of the screen is Ihde's *embodiment relation*,[8] which he describes as a particular kind of body-technology coupling characterized by the transformation of our perception—where the technology "enters into my bodily, actional, perceptual relationship with my environment."[9] The screen-body schemata thus enacts particular ways of seeing and knowing, while the medium of the screen exhibits a specific intentionality or telic inclination; for example, as we will later suggest, the "framing of relevance," the demand for (audio)visual attention, is one of the intrinsic affordances of "screen-ness" in Introna and Ilharco's terms.[10]

We can see, then, how the material and perceptual specificity of media interfaces and apparatuses are deeply integral to our individual and collectively realized corporeal schemata. Our corporeal schemata are in an emergent and dynamic *relation* with our environment and its affordances—always a mode of doing, always situated and contextual, and always a more-or-less partial inflection of our collective and culturally specific habits. Casey's adaptation of Pierre Bourdieu's notion of *habitus* provides some insight into the relation between collective embodiment and place that can be usefully applied to the body-screen coupling. For Casey, *habitus* is the "mediatrix of place and self,"[11] and the body and its flexible schemata is the vehicle that enacts both routine and improvisation within the place-world. He writes: "A given *habitus* is always enacted in a particular place and incorporates the features inherent in previous such places, all of which are linked by a habitudinal bond."[12] Thus, places invoke our conceptual, somatic and "felt" memories and familiarities, and *habitus* is the dynamic mergence of this "routine" with the complex contingencies of our local and material environment. Casey's habitus has some correlation with the phenomenological concept of intercorporeality, as both grasp the way individual bodies digest the collective embodiments or shared habits of their wider cultural milieu. As we will suggest, windows and frames are durable and sedimented habitudinal encounters that impact upon our experience of place and emplacement; more significantly, there is a robust habitudinal bond between windows, frames, screens and bodies that is integral to our contemporary intercorporeality.

Post-phenomenology is an apt method of analysis as it counters the notion that disembodiment is a condition of using the Internet or the phone, is able to consider the ways that tele-technologies modify the body, and the kind of embodiment afforded by telepresent and mobile media. As Adriana de Souza e Silva and Daniel Sutko note, early discussions about cyberspace and the Internet distinguished between the virtual and "real" world, resulting in a "conceptualization of the Internet and digital information as essentially disconnected from physical spaces."[13] Mobile devices clearly antagonize any notion of a disembodied telepresence that is seemingly endemic to digital screen media, as we are frequently on-the-move, on-the-street and purposefully situated in local spaces and places when engaged in mobile phone use and mobile gameplay. Moreover, as de Souza e Silva suggests, location-aware mobile media and the increasing use of navigational and/or place-specific applications effectively interweave the physical and the digital, dissembling the dualism as both come together in "the immanence of the real."[14] For instance, Gordon and de Souza e Silva (whose work we shall return to later) argue that "location-aware technologies allow us to seamlessly interact with the remote and the proximate" in ways that complicate attempts to draw clear delineations between "co-presence" and "networked interaction."[15] Intermingling of this kind is evidenced in many different developments, from (among other things) augmented reality applications, to the wide array of "geomobile" applications which, Crawford and Goggin suggest, are "transforming the relationship between data space and physical space [and place]."[16] One way of understanding this interweaving of the physical with the digital is as a meshing of located place and networked space—a formulation which raises a number of rich questions. Do mobile media work to collapse the space-place distinction, or enable "space" and "place" to be simultaneously present? Or, does the corporeal schema "dilate" (in Merleau-Ponty's sense) to apprehend virtual and screen spaces as "places."

Conceiving of these developments as suggestive of a meshing of located place and networked space presents difficulties for certain theorizations of the space-place relationship. For instance, according to Casey, space is an abstraction of place, while place is the material "actuality;" that is, space doesn't demand the "full engagement" that place does, and the virtual doesn't have "habitudinal density."[17] Yet, mobile media problematize this "polar ontology," merging or embedding the experience of (abstract, "virtual") screen space and actual place (particularly in pedestrian contexts), while the hands-eyes-screen relation merge embodied "place" or "habitus" with screen space (particularly in the case of touch screens and mobile gaming). For Casey, places exist within or alongside one another in space, yet frames and screens—particularly small mobile ones—reverse that relation. That is, when we're playing a game or searching the web, we enact and perceive "space in the hand." In the case of mobile screens, then, we need a different way of thinking through the space-place relation; what modalities of habitude, anchoring, dwelling and movement are particular to mobile screen experience? The mobile screen/frame itself provides a kind of anchoring, but the mode of attachment differs depending on the activity. Thus, relations between users and their mobile media devices reveal a number of modalities of presence: located presence, co-presence and telepresence. For example, the way in which the mobile phone as casual game device is used to effect a mode of non-communicative co-presence when one is "alone" in public, a modality quite distinct from the much-discussed telepresent effects of talking or texting on the mobile, or from our telepresent immersive experience of first and second screens (TV and computer). So casual gaming can be seen as a mode of co-present telepresence, wherein one is simultaneously with-others and in-the-game. Thus, mobile media use occurs across a spectrum of "placing" and "presencing."

Casey argues that glocalization (such as that which we experience on the Internet, the linking of every locale to every other in global space) results in the "thinning out" of places such that they become more spatial. That is, he argues, screens compromise the integrity of place. In fact,

> they do not even hold, lacking the rigor and substance of thickly lived places . . . Their very surface is perforated, open to continual reshaping and connecting with other surfaces. Think of the way in which programs on television or items on the Web melt away into each other as we switch channels or surf at leisure. In such circumstances, there is a notable *lability of place* that corresponds to a fickle self who seeks to be entertained.[18]

This distracted self, then, is concurrent with the instability of screen surfaces that cannot hold places within them. Yet we often conceive of mobile phones as containers—"holders" of personal information, digital content, applications, game worlds—and we have to re-think the ontology of mobile screens as an aspect of that containment. Moreover, the embodiment relation we have with mobiles is particularly complex, both in terms of their near-constant closeness to the body, and increasingly, the way they "enter into" our actional and perceptual experience of the world, resulting in a perceptually hybrid experience of place that merges online and physical environments (such as is the case with navigational, mapping, and place-based applications). To develop this further, it is valuable to ask the following: If, as Casey suggests, the body is the "enactive vehicle" of the place-world, enabling a sense of place or "here-ness,"[19] how do mobile screens appear in that place-world and what is their involvement in the body-place relation? How does the body incorporate mobile screens into its corporeal schema? And, what is the corporeal/embodied effect of this screen upon our perception of place?

A productive means of getting a purchase on these issues—in particular, as a way of understanding the perceptual effect of mobile screens and their transgressive "boundary work"—is by considering one of the more common metaphors of the screen—that of the frame, or "window-on-the-world," and other forms of what Casey would term habitudinal "edge phenomena."[20] In other words, one way that we can begin to rethink screens as places or place-holders is by examining what "frames" our use of mobile screen media in place, as well as by examining the mobile device as itself a kind of "frame" that mediates outside/inside dynamics. That is, the mobile as embedded in an "outside" of body-technology relations *in place*, as well as a technology with an "inside" in the sense of acting as a "container" for various forms of mobile media content. These dimensions will be explored in the following section.

FRAMES AND EDGE PHENOMENA

The ontological and cultural significance of the window and the frame, especially for understandings of media, cannot be overstated; as Anne Friedberg comments, the frame is perceived as "the decisive structure of what is at stake,"[21] while for Vivian Sobchack it is both a "lived logic" and itself "an organ of perception."[22] The comparison between screen and window as framing devices is easily made and understood—the frames of window and screen are similarly rectangular, they can be similarly interpreted as membranes between "inside" and "outside," and what one sees through the frame is a portion of the world in space and time (in the case of television, for example, this is often aimed towards a "realistic" depiction of a place/event in a parody of the scene-through-a-window).

It is worth examining in some detail the portrayal of the screen as frame or window, in the first instance, especially, in relation to traditionally less-mobile forms of screenic media, such as film and television, prior to an examination of the screen as frame or window in the context of mobile media.

In technosomatic terms, traditionally, we might consider the window, frame and screen as perceptually inter-familiar, exhibiting a kind of onto-logical consonance; that is, as Friedberg notes, like the window, the screen and its frame "holds a view in place;" it becomes, like the window, a trans-formative aperture in architectural space, altering the materiality of our built environment and opening surfaces up to a new kind of conceptual and metaphoric "ventilation."[23] The screen-as-window, then, sets up a par-ticular kind of corporeal trope: to look out a window and to view a screen, at the imperative of the eyes and face, one's body must be turned towards the apparatus. As such, to remain visually attached the body is rendered immobile. Indeed, for Lev Manovich, this fixedness typifies a bodily inertia and sensory deprivation that has been and remains a predisposition of "the Western screen-based apparatus" in general.[24] This tendency can be traced from Alberti's perspectival window and Renaissance monocular perspec-tive, through to Kepler's *camera obscura*, nineteenth century *camera lucida* and contemporary cinema: in all of these interfaces, Manovich argues, the body is fixed in space.[25] Although the dynamic screens of cinema and tele-vision might be said to virtually transport the viewer, Manovich argues that this mobility is had at the cost of the "institutionalised immobility" of the body of the spectator,[26] in the form of the silent seated rows of movie-goers or the domestic couch-reclining TV viewer. Importantly, this distinc-tion between the virtually mobile or tele-active eye-body and the stationary physical body is made by way of the screen-as-window metaphor. Within this metaphor the eyes alone must remain mobile, to traverse and visually "handle" the surface space of the screen, while the face and body are held captive by the eyes' attachment. The rectilinear dimensions of the media window—and its immobilization of the body in front of the screen—is an instance of the epistemological containment of knowledge in perspectival vision, today most familiar through the ubiquitous frame of the screen.

Introna and Ilharco extend this understanding of perspectival vision in their articulation of some of the more general ontological properties of what they term "screen-ness." In contemporary life, screens are often a pri-mary focus of our attention and concern: they literally display that which is relevant or worthy of notice. This property of relevance has little to do with the specific content of any particular screen display; rather, it indicates, in the words of Introna and Ilharco:

> a particular involvement in-the-world in which we dwell and within which screens come to be screens. It is not up to anyone of us to decide on the already presumed relevance of screens; that is what a screen is—a framing of relevance, a call for attention, a making apparent of a way of living.[27]

Introna and Ilharco suggest that screens of all kinds enter our involvement-in-the-world at the moment we turn them on, at which point we reposition our attention and "sit down, quit—physically or cognitively—other activities we may have been performing, and watch the screen."[28]

Both of the previous cases—the arguments presented by Manovich, and those by Introna and Ilharco—develop a particular understanding of body-screen-place relations. Yet, as we will go on to argue, this "frontal" relationship which is typical of our engagement with most screens—where the media of cinema, television and computer can be said to discipline the body more or less into a face-to-face interaction—is thoroughly "disintegrated" by the mobile media screen. Our interaction with mobile screens is rarely marked by such dedicated attentiveness; indeed, our "turning towards" them is usually momentary (checking for a text or missed call) or at most can be measured in minutes. In other words, our engagements with mobile screens lead to a distinctly different set of body-screen-place relations which implicate both how mobile screens appear in the place-world, and the corporeal/embodied effects of the screen upon our perception of place.

In order to tease out how this is the case, it is valuable to consider Derrida's quite specific understanding of frames, especially as developed in relation to the notion of the *parergon*. In *The Truth in Painting*, for instance, Derrida explores the idea of the *parergon* in relation to two passages in separate texts by Kant.[29] The Greek word *parergon* literally means "outside the work" as *par* means "past" or "beyond," and *ergon* means "work." Contrary to the suggestion of clear separation, however, the implication is that there is a close and mutually influencing proximity between *ergon* and *parergon*—this is in part because the prefix *par* also means "beside." In this way, the *parergon* "does not fall to one side" of the *ergon*, rather "it touches and cooperates, from a certain outside" which is "neither simply outside nor simply inside."[30] As Derrida goes on to explain: the *parergon* "comes to play, abut onto, brush against, rub, press against the limit itself and intervene in the inside only to the extent that the inside is lacking."[31] This is the case insofar as the inside—the "work"—"is not complete, exhausted;" rather, "it needs this supplement [the *parergon*] to finish it."[32]

In working through the implications of this relationship between *ergon* and *parergon*, Derrida explores three examples: the status of clothing on statuary; the role of columns on buildings; and, frames for paintings. In the present context, it is the last of these that is most instructive for understanding body-*mobile screen*-place relations, and our phenomenological engagements with mobile handsets as a "container" for mobile media content. According to Derrida, in an art context, a painting's frame as *parergon* serves two vital yet seemingly contradictory simultaneous operations. On the one hand, its precise function is to stand out "like a figure on a ground," both from "the ergon (the work)" and also from "the milieu" (which in the case of displayed art, includes both the immediate surroundings of the gallery wall, *and* a wider art context).[33] This it does in order to "delimit" the content of any given work. On the other hand, and at the same time, "the *parergon* is a form which has as its traditional determination not that it stands out but that it disappears, buries itself, effaces itself, melts away at the moment it deploys its greatest energy."[34] In this sense, then, Derrida's conception of the frame is closer in meaning to the concept of the boundary than that of the border. As Casey explains, "in contrast with borders, boundaries are permeable; they are porous, full of holes; they allow, indeed often invite, movement across them," while, at the same time, remaining "edges of a certain definite sort."[35]

This understanding of the frame as *parergon* is most instructive for thinking through the complex interactions that are involved in our body-place orientations to mobile media. In particular, the mobile screen (and the handset in which it is embedded) serves as the *parergon* that both "frames" the *ergon* of mobile media content but always in such a way that the separation between content and user experience of place ("the milieu") "disappears, . . . effaces itself, melts away."

A key implication of this specific understanding is that it assists us in making sense of what Heidi Rae Cooley has characterized as an increasingly "material experience of vision" by mobile media users, where "hands, eyes, screen, and surroundings interact and blend in syncopated fashion."[36] To put this another way, oscillating technosomatic registers of attention, inattention and distraction are enacted when engaging with small mobile screens, and such engagement undermines both the facial dedication of the immobilized body deemed typical of our embodied relation to larger screens, and consequently the frontal ontology of window and the frame (contra Manovich). What is more, our relationship with the mobile phone as a multi-sensory device which can be used either as a dedicated aural *or* visual medium, for example, can effectively shift eye-behavior from a continual fixed-ness on the screen to a sporadic, oscillating and context-dependent mode of viewing (contra Introna and Ilharco). In such circumstances, the dedicated frontal orientation we have towards screens becomes compromised by our own mobility, the screen size and resolution, and the interrupted nature of mobile phone use.

In the remainder of this chapter, we draw together the various threads of our argument—how mobile media suggest certain body-technology reorientations, the interrupted nature of mobile phone use, how these media affect our experiences of place, and how they cross the "boundary" of the screen frame—in two separate examples or case studies. The first concerns web-based mobile practices, and the second sound-based mobile technologies.

WEB-BASED MOBILE PRACTICES: LOCATION-BASED APPLICATIONS, THE "UNCONTAINABLE" METAPHOR OF THE NETWORK, AND "NET-LOCAL PUBLIC SPACE/PLACE"

De Souza e Silva and Frith note that both the GPS-enabled iPhone and Google's Android operating system "contributed to the popularization and commercialization of location-aware applications" and location-based services, which typically provide situational information about the urban environment via online databases and media libraries, such that informational changes on the mobile screen effect the navigation and experience of physical space.[37] Interestingly, West and Mace point out in their analysis of the iPhone's market success, that the increase in mobile Internet use was a result of an important shift in thinking: from the provision of an Internet tailored to mobile screens (e.g. evidenced by micro-browsers and provider portals) to the provision of the "real Internet" on the mobile device, initiated with the iPhone. They state:

> Three weeks before the release of the first iPhone, Steve Jobs predicted, "people want the real Internet on their phone" . . . Jobs was eventually proven correct: when given web browsing that was substantially similar to the browsing experience on a PC, mobile web usage went up dramatically. The success of the iPhone demonstrated that what was holding back demand for mobile data services in the US was not the creation of new mobile-specific value networks, but the delivery of devices and networks that were capable of delivering a convincing approximation of the familiar wired Internet.[38]

Although this suggests the dominance and persistence of desktop-based Internet imaginaries, a number of theorists have argued that the proliferation of mobile online activities—via mobile phones, laptops, personal digital assistants (PDAs), and more recently, netbooks and iPads—are changing the way we think about being "on," "at" or "in" a simulated or computer space, and the way we think about being "on" or "off" line.[39] As de Souza e Silva claims: "Because many mobile devices are constantly connected to the Internet . . . users do not perceive physical and digital spaces as separate entities and do not have the feeling of 'entering' the Internet, or being immersed in digital spaces, as was generally the case when one needed to sit down in front of a computer screen and dial a connection."[40] Being online and networked thus becomes another function of the mobile phone, but it is importantly a *different* experience of the Internet and online connectivity: the supposedly dematerializing effects of cyberspace and telepresent interaction become enfolded inside present contexts and activities, like the embodied and itinerant acts of walking, driving, face-to-face communication and numerous other material and somatic involvements.[41]

Indeed, de Souza e Silva suggests that the increase in web-based mobile practices within urban space means that we need to rethink the spatial and place-based experience of being-in-public. That is, the use of location-based applications generate hybrid spaces by integrating online information of one's immediate environment into the patterns of urban life and peripatetic movement. In recent work, de Souza e Silva and Gordon have argued that such hybrid practices generate what they term *net-local public space*, which describes our movement "between the immediately proximate and the mediately distant within a carefully crafted set of social rituals that ultimately serves to extend the purview of local space."[42] In Ihde's terms, this describes a particular body-technology relation effected by web-capable mobile phones, such that the affordances of the device are taken into our contemporary and collective spatial experience and we begin to perceive "through the reflexive transformation of [our] perceptual and body sense."[43] Net-local public space includes those engaging in location-based activities with mobile devices, those (both co-present and online) inhering or participating in this network activity, and those non-participants who are co-located in the urban setting. As our "attentional foci"[44] in such quotidian spaces become diversified and hybridized, effecting new micro- and macro-movements, our cultural and collective sensoria shifts, and the actual/virtual dichotomy previously used to differentiate between offline/ online practices is thoroughly disassembled into a complex and dynamic range of modalities of presence.

In his phenomenological study of the spatial experience of telephone use, Rosenberger usefully deploys Ihde's notion of the embodiment relation, and adds another variable: *field composition*. This term describes "a technology's potential to reorganize the overall structure of one's field of awareness as the technology is used."[45] Borrowing from Gurwitsch, he suggests that as a technology becomes embodied, we experience a change to our *theme* (that which at any given moment is of central concern or a focus of attention), our *thematic field* (the experiential context of the theme), and that which resides at the margins or withdraws from awareness. This reorganization, Rosenberger argues, "involves a specific relation to the immediate space surrounding the user's body,"[46] but also reorganizes our perception and experience of the broader situation and the boundary between awareness and non-awareness. Rosenberger's approach also accords in key respects with other phenomenologically inspired approaches to the study of place, such as that by David Seamon, who writes of "time-space routines" which "coalesce" to form a "place-ballet."[47] Seamon's argument, as Tim Cresswell summarizes it, is that "places exhibit a kind of unchoreographed yet ordered practice that makes the place."[48] What both approaches share is a common understanding that experiences become conceptually and perceptually "sedimented" with habitual or routine use—and, as de Souza e Silva would suggest, such a sedimentation is occurring with the emergence of net-local public space.

A number of theorists have shown location-based mobile gaming as a particularly robust example of the emergent hybrid ontology or "field composition" of mobile media (de Souza e Silva, Farman, Follet, Copier). In his analysis of Geocaching[49] Jason Farman describes the mixed or augmented realities of pervasive location-based games where bodies, networks and material space converge.[50] In such games, our embodied proprioception— that is, the awareness of our body's position in relation to the environment enabled by our perception of movement and spatial orientation—must seamlessly accommodate both immediate and mediate being-in-the-world. Players of location-based games and users of location-based services, he comments, navigate the landscape in a "simultaneous process of sensorial movement through streets and buildings and an embodied connection to how those places are augmented by digital information on mobile devices." This mergence increasingly constitutes the "interface of everyday life."[51]

SOUND-BASED TECHNOLOGIES: THE IMPACT OF MOBILES ON THE "SOUNDSCAPE" OF PLACE, PLUS BOUNDARY TRANSGRESSION (AND MAINTENANCE)

As Michael Bull argues, portable sound-based technologies such as the Walkman, mobile phone, and mp3 player (such as the iPod) have contributed, along with the automobile, to the transformation of the urban soundscape by way of an auditory privatization of public space.[52] That is, such devices allow the user to control sound—"to manage and orchestrate their spaces of habitation." While the iPod or mp3 player provides a continuous sound-bubble or "sonorous envelope" that effectively allows the user to deny the contingencies of the outside world,[53] the mobile phone is experientially discontinuous, "puncturing" the soundscape via the sporadic and unpredictable contingency of unexpected calls and text messages. The mobile music player is thus discrete and cocooned, whereas the mobile phone user "colonizes" urban space, intermittently carving out a place of communication and telepresent intimacy, temporarily irrupting the immediate soundscape with personal ringtones, bleeps and one-sided conversations. This colonization often requires a complex negotiation of public and private physical place and auditory space; our pedestrian trajectories can be quite radically revised and re-possibilized by the interruption of a mobile phone call or text message, by the beep of one's PDA warning of an impending meeting or deadline, or by those telepresent on the other end of the phone becoming "virtually" integrated into or effecting a change to one's route ("can you pick up some milk on your way?").

Nevertheless, it is the case that both music players and mobile phones are transforming (and continue to transform) our co-proximate or co-present behavior in public places, and the way we inhabit and negotiate urban spaces. Indeed, Bull insightfully observes that in mobile privatization there is a desire for proximity, for "mediated presence that shrinks space into something manageable and habitable."[54] The mobile phone in particular offers us the possibility of proximity with familiar others while on-the-move, and frequently this is via an aural communicative telepresence (though this is complexified by video-phoning, image/text messaging, and increasingly, the proliferation of web-based made-for-mobile applications). There is no doubt that digital media and communication technologies have irretrievably altered our normalized sense of embodied "location" and "presence." As numerous theorists and philosophers have suggested over the past decade or so, our increasing remote interaction with the world—the possibility of extended intervals of telepresence or telematic perception—indicates a need to rearticulate our collective embodiments, to think through other ontologies, other ways of being-in-the-world, and in a Heideggerian sense, of being-with-equipment.[55] In general, the body-telemedia relation modifies spatial and sensory perception, by changing what is "proximal," or the relation between "here" and "there," into a kind of "distant presence" that becomes part of the "as if" structure of our awareness (*pace* Heidegger).

In their study of listening, corporeality and presence in real and artificial environments, Turner et al. comment on the unique qualities of auditory space.[56] Unlike pictorial or televisual space, they argue, auditory space is not "boxed-in" or framed as a window-on-the-world; instead, it is fluid and dynamic, and filtered rather than contained or "stopped" by material obstacles such as walls and corners. Thus, it is not enclosed or "held" in place, but creates "its own dimensions moment by moment."[57] It is this elasticity and pervasiveness ascribable to sound that in turn educed a particular way of experiencing the audile telepresence and sense perception of landline telephony. As it came into common usage, the early landline telephone, Connor argues, allowed our interior body—the inner ear—to be pervaded *"almost without mediation"* by the "vocal body of the other."[58] Thus, with telephonic telepresence we were afforded the impression of being corporeally tethered to another's body in real time, and thus paradoxically co-present at-a-distance—what Backhaus refers to as "the modified we-relationship" of telephone space.[59] Yet, although the fixed landline telephone was one of our earliest experiences of what Helyer calls "schizophonic" audio, allowing us as it were to hear non-present voices, it was still the case that each correspondent was fixed in a particular geo-spatial location—so "the telephonic act became a sonic bridge between familiar sites."[60] With the mobile phone, because the location of the caller is often unknown, shifting, or unpredictable, communication becomes "de-territorialized." That is, the particular consequence of the mobile phone upon aural telepresence was an unfixing from place—effecting the emergence of a kind of *nomadic telepresence*[61] that is only partially overcome by asking that most common of questions in mobile communication—"Where are you?" Paradoxically, or perhaps in spite of, this unhinging of one's geo-spatial location, the itinerant attributes of mobile phone conversation often results in a sense of discomfiture or ill fit between the conversation-context and the very public spaces (buses, trains, sidewalks, etc.) wherein we find ourselves "on the mobile." This awkwardness is perhaps a result of engaging in telepresent communication in public places, of having to negotiate between "presents" (that is, both telepresence and co-presence), because mobile phone communication when one is on-the-move and on-the-street is rarely immersive and private, but demands an awareness of both others' and one's immediate surroundings. Thus, mobile media require us to adeptly shift between actual and telepresent space and place; to "behave" in ways that accord with (or deviate from) conventional modes of being-on-the-phone. The body thus undergoes aural and spatial disciplining required by the device, the collective mobile-user habits of a culture, and the demands of telepresent communication in public places.

CONCLUSION

In this chapter, we have argued that our engagement with mobile screen media changes our experience of place and presence. Mobile media devices, we have suggested, call for a distinctly different set of body-screen-place relations than those of many other forms of (especially fixed or largely stationary) screen-based media. In developing this argument, we have anchored our investigation in post-phenomenology. The advantage of this approach is that it draws attention to certain specific body-technology reorientations that are crucial to understanding contemporary developments such as mobile media technologies. Post-phenomenology does this by giving particular emphasis to how the material and perceptual specificity of media interfaces and applications are deeply integral to our embodied individual and collectively realized corporeal schema. Utilizing the insights generated from this method, this examination explored the ontological and cultural significance of the frame as a way of drawing out how mobile screen media are both embedded in an "outside" of body-technology relations in place, as well as functioning as a technology with an "inside" in the sense of acting as a container or "place-holder," thereby countering arguments that screens cannot contain or hold places within them.

Through the two case studies, we have also explored how mobile use crosses a spectrum of "placing" and "presencing," and sought to demonstrate how interruption and distraction is a key characteristic of our engagements with mobile screen media. The first case study explored how, through hybrid practices, our "attentional foci" become diversified and hybridized, affecting a whole range of bodily and cultural shifts that are ultimately disassembled into a complex and dynamic range of differing modalities of presence. In this sense, our experiences with and of the mobile become perceptually "sedimented" with habitual use, and increasingly infused with a sense of networked and net-local place. Through use of mobiles and various location-based gaming services, our engagements with urban place become hybridized to such an extent that they realize in completely routine ways that which McKenzie Wark has referred to as the overlay of the "third nature" of electronic communications over the "second nature" of the built environment.[62]

The second case study explored how colonization of urban space by iPods, PDAs, and other mobile technologies requires a complex negotiation of public and private physical place, auditory space and pedestrian trajectories. These work to transform co-proximate encounters and a "normalized" sense of embodied location. Both studies showed how we negotiate a number of overlapping "presents" (telepresence, co-presence, located presence and net-local presence), and described how the body undergoes aural and spatial disciplining as a combined result of our engagements with the mobile device, the collective cultural habits associated with mobile use, and the demands of telepresent communication in public places.

Finally, the material of these two cases illustrates that the physical, perceptual, and contextual mode of our being-in-the-world is phenomenologically altered by mobile engagement. Yet, as we have suggested, a sense of place—if somewhat altered—remains integral to the body-mobile relation, as it is both "contained" in the micro worldliness of the device and hybridized by the perceptual merger of situated "hereness" with online networks.

NOTES

1. Jacques Derrida, *The Truth in Painting*, trans. Geoff Bennington and Ian McLeod (Chicago, IL: The University of Chicago Press, 1987), 57.
2. Tim Cresswell, "Place" (2009), http://www.elsevierdirect.com/brochures/hugy/SampleContent/Place.pdf.
3. Anne Friedberg, *The Virtual Window: From Alberti to Microsoft* (Cambridge, MA: MIT Press, 2006).
4. Ingrid Richardson and Rowan Wilken, "Haptic Vision, Footwork, Place-making: A Peripatetic Phenomenology of the Mobile Phone Pedestrian," *Second Nature: International Journal of Creative Media*, 2(1) (2009), http://secondnature.rmit.edu.au/index.php/2ndnature/article/view/121/35.
5. Maurice Merleau-Ponty, *The Primacy of Perception, And Other Essays on Phenomenological Psychology, the Philosophy of Art, History and Politics* (Illinois: Northwestern University Press, 1964), 5.
6. Don Ihde, *Technics and Praxis* (Dordrecht, Holland: D. Reidel Publishing Company, 1979).
7. Don Ihde, *Technics and Praxis,*17.
8. Ihde's other relations include: *hermeneutic* relations (semi-opaque relations which require us to "read" a device such as a thermometer or gauge); *alterity* relations (technology as "other" such as the computer as artificial intelligence); *background* relations (such as the hum of electricity); *horizonal* relations (the "texturing" of the environment effected by technological infrastructure that is at the fringes of embodiment or beyond direct perception). Don Ihde, *Technology and the Lifeworld: From Garden to Earth* (Bloomington, IN: Indiana University Press, 1990), 112–114.
9. Don Ihde, *Postphenomenology and Technoscience: The Peking University Lectures* (New York: SUNY Press, 2009), 42.
10. Lucas D. Introna and Fernando M. Ilharco, "The Ontological Screening of Contemporary Life: A Phenomenological Analysis of Screens," *European Journal of Information Systems*, 13 (2004).
11. Edward S. Casey, "Between Geography and Philosophy: What Does It Mean to Be in the Place-World," *Annals of the Association of American Geographers*, 91(4) (2001), 686.
12. Casey, "Between Geography and Philosophy," 686.

13. Adriana de Souza e Silva and Daniel M. Sutko, "Theorizing Locative Technologies through Philosophies of the Virtual," *Communication Theory*, 21(1) (2011), 25; see also, Rowan Wilken, *Teletechnologies, Place, and Community* (New York: Routledge, 2011), 21–25.
14. de Souza e Silva and Sutko, "Theorizing Locative Technologies," 34.
15. see Eric Gordon and Adriana de Souza e Silva, "The Urban Dynamics of Net Localities: How Mobile and Location-aware Technologies are Transforming Places," this volume.
16. Alice Crawford and Gerard Goggin, "Geomobile Web: Locative Technologies and Mobile Media," *Australian Journal of Communication*, 36(1) (2009), 98.
17. Casey, "Between Geography and Philosophy," 686.
18. Casey, "Between Geography and Philosophy," 684–685.
19. Casey, "Between Geography and Philosophy," 687.
20. Edward S. Casey, "Taking Bachelard from the Instant to the Edge," *Philosophy Today* (Celina, OH: Messenger Press, 2008), 8.
21. Friedberg, *The Virtual Window: From Alberti to Microsoft*, 14.
22. Vivian Sobchack, cited in Friedberg, *The Virtual Window*, 16.
23. Friedberg, *The Virtual Window*, 1.
24. Lev Manovich, *The Language of New Media* (Cambridge, MA: MIT Press, 2001), 104.
25. Manovich, *The Language of New Media*, 104–105.
26. Manovich, *The Language of New Media*, 107.
27. Introna and Ilharco, "The Ontological Screening," 221–234.
28. Introna and Ilharco, "The Ontological Screening," 225.
29. Derrida, *The Truth in Painting*.
30. Derrida, *The Truth in Painting*, 54.
31. Derrida, *The Truth in Painting*, 56.
32. Barbara Mella, "Derrida's Detour," *Reconstruction: Studies in Contemporary Culture*, 2(4), Fall (2002), ¶40, http://reconstruction.eserver.org/024/mella.htm#53A.
33. Derrida, *The Truth in Painting*, 61.
34. Derrida, *The Truth in Painting*, 61.
35. Casey, "Taking Bachelard," 8.
36. Heidi Rae Cooley, "It's All About the *Fit*: The Hand, the Mobile Screenic Device, and Tactile Vision," *Journal of Visual Culture*, 3 (2004): 145.
37. Adriana de Souza e Silva and Jordan Frith, "Locative Mobile Social Networks: Mapping Communication and Location in Urban Spaces," *Mobilities* 5(4) (2010), 486.
38. Joel West and Michael Mace, "Browsing as the Killer App: Explaining the Rapid Success of Apple's iPhone," *Telecommunications Policy* 34(5–6) (2010), 282.
39. Frank Lantz, "Big Games and the Porous Border between the Real and the Mediated," *receiver* 16 (2006), http://www.receiver.vodafone.com/.
40. Adriana de Souza e Silva, "From Cyber to Hybrid: Mobile Technologies as Interfaces of Hybrid Spaces," *Space and Culture*, 9(3) (2006), 263.
41. Though it is not possible to expand on the following point here, the "corporeal turn" in cultural and post-semiotic theory, which has incorporated the work of Merleau-Ponty and other phenomenologists, has been used effectively as a counter-argument against those who claim that online interaction is "disembodied." In this context, the oppositional relation between virtual and locative spaces would be moot, since both dedicated Internet use and use of context-aware mobile devices would be considered different ways of being embodied, each educing distinctive corporeal schematics.

42. Adriana de Souza e Silva and Eric Gordon, "Net-local Public Spaces: Towards a Culture of Location," unpublished paper (2010), http://www.urbancomm. org/dynamic_images/seminars/.
43. Don Ihde, *Technology and the Lifeworld: From Garden to Earth* (Blooming-ton: Indiana University Pressm 1990), 72.
44. de Souza e Silva and Gordon, "Net-local Public Spaces."
45. Robert Rosenberger, "The Spatial Experience of Telephone Use," *Environment, Space, Place* 2(2) (2010), 66.
46. Rosenberger, "The Spatial Experience of Telephone Use," 69.
47. David Seamon, "Body-subject, Time-space Routines, and Place-ballets," in Anne Buttimer and David Seamon (eds), *The Human Experience of Space and Place* (New York: St Martin's Press, 1980), 148–165.
48. Tim Cresswell, "Place," 7.
49. Geocaching is "a worldwide GPS treasure hunt game that is played in over 200 countries . . . The game players hide geocache containers in public areas, marking them with GPS coordinates. Players use their mobile devices (from GPS receivers to iPhones) to track down the container, sign the log, and leave tradable and trackable items in the cache." Jason Farman, "Locative Life: Geocaching, Mobile Gaming, and Embodiment," *Proceedings of the Digital Arts and Culture Conference—After Media: Embodiment and Context* (University of California, Irvine, December 12–15, 2009), http://escholar-ship.org/uc/item/507938rr.
50. Farman, "Locative Life."
51. Farman, "Locative Life."
52. Michael Bull, "Thinking about Sound, Proximity and Distance in Western Experience: The Case of Odysseus's Walkman," in *Hearing Cultures: Essays on Sound, Listening and Modernity*, ed. by Veit Erlmann (Oxford and New York: Berg, 2004).
53. Bull, "Thinking about Sound," 185, 189.
54. Bull, "Thinking about Sound," 177.
55. Martin Heidegger, *The Question Concerning Technology and Other Essays* (New York: Garland Publishing, Inc., 1977).
56. Phil Turner, Susan Turner and Iain McGregor, "Listening, Corporeality and Presence," *Proceedings of the 9th International Workshop on Presence* (London, September 2005), http://citeseerx.ist.psu.edu/viewdoc/download?d oi=10.1.1.92.7329&rep=rep1&type=pdf.
57. Carpenter cited in Turner et al, "Listening, Corporeality and Presence," 44.
58. Steven Connor, "Sound and the Self," in *Hearing History: A Reader*, edited by Mark Michael Smith (Athens, University of Georgia Press, 2004), 56, italics ours.
59. Gary Backhaus, "The Phenomenology of Telephone Space," *Human Studies* 20 (1997), 203.
60. Nigel Helyer, "The Sonic Commons: Embrace or Retreat?" *Scan: Journal of Media Arts Culture*, 4(3) (2007), http://scan.net.au/scan/journal/display. php?journal_id=105.
61. Helyer, "The Sonic Commons."
62. McKenzie Wark, "Third Nature," *Cultural Studies*, 8(1), January (1994), 115–132.

12 Encoding Place
The Politics of Mobile Location Technologies

Gerard Goggin

Technologies of location for cellular mobiles have been in development for some years, since at least the late 1990s. Mobile media are now awash with various kinds of location technologies that have greatly expanded earlier conceptions of how place could be constructed, capitalized on, and mobilized through personal, portable technologies. As well as the location technologies developed by cell phone companies and carriers, we have also seen the rapid development of Bluetooth, whether in advertising or user file sharing; satellite navigation (sat nav) technologies; geoweb applications (Google Earth and Google Maps); the marshalling of location in mixed and alternative reality mobile gaming; finding of friends, intimates, and new contacts with mobile social software; annotation, photographing, filming, recording and mark-up of locales through mobile Internet applications; smartphone apps that take advantage of the possibilities of location technologies. These location technologies are now a vitally important part of the relations between place and mobile technologies, and are now receiving overdue attention and theorization—posing a double-task for such research, and for this chapter also.

In the first place, the shaping, appropriation, construction, and marshalling of place when it comes to mobiles is something that requires further recognition and systematic discussion—as proposed by this volume. Making links between place theory and research on mobiles is an obvious way to embark on this project. It is especially helpful because the work of many theorists represented and discussed in this volume reminds us of the need to carefully attend to the work of making place that has been occurring with mobile technologies, and which now, with the "locational" turn, takes new directions.

Second, the specific systems, affordances, and infrastructures involved in the media ecology of location technologies means that there is a new dimension to this place making, with new power relations, that requires analysis and debate. We might talk about this as the "encoding" of place, to take advantage of the resonances of political analysis and critique of software—and also network infrastructure and devices—undertaken by a range of theorists from quite different positions, including Lawrence

Lessig,[1] Alexander Galloway,[2] Mark Hansen,[3] Ravi Sundaram,[4] Sarai,[5] and others.[6] The aspiration of this chapter, then, is to contribute to the understanding of mobile emplacement by focusing on this now prominent encoding of place—with a particular focus on cellular mobile technologies.

We now have exciting research emerging regarding the politics of location technology and its relationship to place. An important account is offered by Eric Gordon and Adriana de Souza e Silva in their 2011 book, *Net Locality*.[7] Their notion of *networked locality* (or net locality)—discussed in their contribution to this volume also—is what they see as the cultural and technological framework through which people manufacture places mediated by location-aware and mobile devices and digital networks. Gordon and de Souza e Silva are especially interested in the relationships that such net locality creates in public spaces, following in the grand tradition of theorizing urbanity, interactions, and publics. In a later book with Jordan Frith, de Souza e Silva develops an argument concerning "mobile technologies as interfaces to public spaces," arguing that, "location-aware interfaces represent important new ways people can filter public spaces." De Souza e Silva and Frith argue that:

> With the increasing use of location-based services, location acquires relevancy as it fundamentally shapes social, political, and spatial interactions . . . Previously understood as a place without meaning, location-based information attached to locations now contribute to construct the meaning of places.[8]

Their arguments are especially animated by a thoughtful, persuasive consideration of the relationships of information, privacy, and power entailed in mobile social media location technologies appearing from 2009 onward in the US in particular, such as Foursquare, Loopt, Brightkite, Latitude, and Whrrl. While inspired by their work, my focus in this chapter is different. As well as engaging with the politics of new mobile social media and what it represents for our accounts of place, I am interested in the interplay and attachments among technologies with distinctly different provenances—yet which now come together to constitute our contemporary mobile media.

Accordingly, this chapter is organized into three main sections. First, I look at the development of location-based technologies, from early efforts by mobile companies, through global positioning system (GPS)-based satnav devices, to the rise of Internet-based "geospatial" web technologies (the most famous being Google Maps and Google Earth). These evolving mobiles and Internet technologies have overlapped considerably, forming distinctive mobile media ecologies of places that are still unfolding—something I illustrate in the second section, through a discussion of place and smartphones. The 2011 outcry over the discovery that Apple kept records of iPhone users' location data—and that these records were easily hacked—is something that highlights the stakes in this emerging phenomenon of encoding place. Accordingly, in the third section, I look at the political and cultural economies of location-based information. I argue that the social relations of how place is produced—indeed encoded—is something that urgently needs scrutiny and debate.

MAPPING MOBILES

Broadly speaking, there are three positioning technologies for mobiles: mobile-based solutions, where positioning is carried out in the handset and sent back to the network; mobile-assisted solutions, where location measurements are relayed to the network, which calculates a position; and network-based solutions, where positioning is done by the network.[9] The cellular mobile industry has been keenly interested in location-based services for many years, especially since governments and regulators, such as the US Federal Communications Commission, stipulated that carriers were obliged to provide emergency services dispatchers with the telephone number of an emergency caller and the location of the cell site or base station transmitting the call.[10] Such moves by policymakers provided additional impetus for cell phone operators to develop location technologies and capabilities for commercial purposes:

> Location-based services provide customers with a possibility to get information based on their location . . . Location services are added value services that depend on a mobile user's geographic position.[11]

Early mobile location-based services centered on providing information to users, trigged by information regarding their location:

> Weather forecasts, tourist attractions, landmarks, restaurants, gas stations, repair shops, ATM locations, theatres, public transportation options (including schedules) are some examples of information provision filtered to the user location.[12]

Such services were relatively slow to develop, though much prototyping and experimentation abounded.

One class of services that did draw on location capabilities of mobiles, but broadened the circle of users involved, involved the use of Bluetooth for advertising, promotion, and user interaction with environments. Bluetooth is a short range wireless connectivity standard, introduced in the late 1990s,[13] with version 2 appearing in 2004, and version 4 now under development. Bluetooth was incorporated in many devices, including laptop computers and mobiles.[14] Bluetooth transmitters could also relatively easily and affordably be incorporated in particular locations, such as advertising billboards.[15] This allowed mobile phone users in the vicinity to be contacted when their Bluetooth was sensed by the billboard. Users were asked if they wished to download a file, which could range from an image to a piece of music (a common promotion in urban environments, or festivals). Users had already cottoned onto the interactional possibilities of Bluetooth, using its ability to detect devices within range in a location to meet others, swap files (especially music), or play games. This resembled the technology for supporting sociability represented in mobile social software ("mososo")—such as the Japanese LoveGety device and US Dodgeball application[16]—that from the late 1990s onward prefigured contemporary mobile social media's fascination with location.

Alongside the development of early mobile location-based services, another class of devices developed during the 2000s that would prove vastly influential. This was the sat nav device that used GPS technology in conjunction with a database of information taken from street directory publishers to provide in-car navigation for drivers. Initially, sat navs were built into cars—of the estimated three million sat navs in US cars in 2005, half were in-car devices.[17] Sat navs were developed by leading companies such as Garmin (founded in 1989) and TomTom (established in 1991). Within a few years, portable, personal navigation devices were replacing street directories, and potentially integrating with smartphones and PDAs.[18] By 2009, there were some thirty million sat nav devices in the US, with over 90 percent of them portable—typically affixed to the dashboard or windscreen.[19]

Sat navs are a prosaic yet fascinating instance of the rise of mapping and location technologies. There was strong interest in using the capabilities of GPS, a pervasive network of satellites first established by the military, for civilian, commercial, and non-commercial purposes.[20] In addition, in the late 1990s and early 2000s, there were many other efforts to develop and commercialize location-based services, especially capitalizing on the growing ubiquity and popularity of cellular mobile phones—but also their potential for providing usable information on the location of the device and its user. Mobile phones also began to incorporate GPS technology, meaning that there was the possibility of triangulating location through both these satellite position technologies and also mobile telecommunications networks. Yet, it was the sat nav that really gained the first solid acceptance. Very likely this was due to consumer expectations of maps in the form of street directories; and consumer needs and desires, in terms of navigation and wayfinding in cars, and the difficulties of doing this when obliged to read a map or directory at the same time as driving.

Thus, sat navs offered a distinctive—and eventually portable—technology, by which the car was connected to networks and databases, and was able to present required information in audio as well as textual and graphical form, to drivers and passengers. In doing so, sat navs become perhaps the first widely used in-car mobile data device—more so than mobile phones themselves (although these were being widely used for texting and perhaps some mobile Internet at the time that sat navs were gaining in popularity). Interestingly, though sat navs were generally used solely for the purpose of navigation, few sat nav devices designed for cars, for instance, sported Internet capability. With the popularity of mobile apps on smartphones, companies like Garmin offered Opencaching—the official geocaching app—and Garmin Mechanic, an app to track car mileage and performance.

Once sat navs become popular, and relatively more affordable, establishing the profitability of the dedicated companies that specialized in them, then mobile phone companies were quick to mount a competitive response. The obvious path for mobile phone companies to take, especially with the advent of multimedia and smartphones—as well as the fact that the mobile devices already had GPS built in—was to modify the software, and supplant the sat nav devices. Thus, a number of mobile phone manufacturers— perhaps most prominently, Nokia—made a concerted pitch to take a share of the domestic sat nav market. Many drivers and passengers were happy to avail themselves of a suitably equipped mobile phone, which they could also mount in their car if they wished, or simply use as desired, to navigate their journey, rather than incurring the cost—and potential inconvenience—of purchasing another device. It is fair to say, however, that the moves by the mobile phone vendors to annex the sat nav market via software extensions were nowhere near as successful as something else that was neither so obvious or predictable: the rise and rise of Google Maps.

Google Maps was announced by the company in 2005, developed first by a Sydney-based company.[21] In 2006, road maps were added, with the multiple directions to destination features following in late 2006.[22] Google Earth, its companion program, which also works in a browser, was publicly released in mid-2005.[23] Google released a Maps for Mobiles application in 2006, upgraded to Google Maps for Mobile 2.0 in late 2007—which featured a "My Location" capability, which can use the GPS information of a mobile device in addition to software that uses information on the location of the nearest mobile transmitter (as it does similarly with WiFi transmitter information from a database).[24] In late 2009, Google released its Maps Navigation app exclusively for Android 2.0 phones.[25] This is a free application with voice prompts, directions, and other features of sat nav devices. Google's Maps Navigation app competes directly with the sat nav devices for cars. However, the sat nav devices are still proving popular— and the sat nav companies also provide apps for mobile devices (Garmin Mobile), and have even entered the mobile phone market themselves (e.g., Garmin's joint venture with Asus to offer the Garminfone, built on the Android platform).

In this section of the chapter, then, I have discussed the development of location technologies as they have been developed for cellular mobile devices, or, having been developed for other purposes, become influential in shaping these. Explicit mobile location-based devices were slow to attract consumer interest, despite the strong commercial motivation by their backers, and others wishing to incubate mobile-based commerce, to gain their acceptance. A similar fate awaited mobile social software, with early experiments failing to become a wider media form. Instead, the development of place-making and referencing technologies in mobiles came from other directions—from the incorporation of Bluetooth into mobiles, the parallel development of sat-nav devices, and, attached to the Internet, the rise of the geospatial web, and mapping applications. This has immediate and significant implications for how we understand mobile technologies in relation to location—and, by extension, to place. From their inception the kinds of location technologies in mobile media—that skew it as locative media—very much work in a media ecology, rather than just being based on a particular technological system. The encoding of place that such mobile technologies enact is a complex affair. Its policy implications can no longer be grasped, for instance, by recourse to the mobiles industry—as was the case with the incorporation of location information in emergency number calling, as in the FCC's early proceedings on E-911. Rather, the place-making which mobile networks, devices, and applications condition and support involve a combination of infrastructures, affordances, and collection of data concerning location. To explore this richly ambiguous encoding, let us turn to the treatment of place with the latest phase of mobile media—the phenomenon of smartphones.

PLACE AND THE SMARTPHONE

In a 2009 paper on the "geomobile web," Alice Crawford and I argued that:

> Underway is large-scale commercial investment joining up sat nav, directories, mapping, dynamic data, and Internet, into new systems for mobile devices. While there is palpable ferment in this field, and it is true that these new, mobile-mediated inventions of place are still at a relatively early stage, important features of these systems are being put in place with little discussion from interested groups. The social and cultural implications of these geomobile technologies, the privacy issues raised by the enormous new quantity of personalized location data, not to mention the kind of networked architectures and environments, and digital spaces that are created—are all something that needs much considered attention and more debate.[26]

In a brief three years, this has now changed. Mobile location technologies are now much more widely used, with take-up slowly but steadily increasing.

For instance, research by the British regulator Ofcom in February–March 2008 found that "11% of mobile phone owners are using the device to access the internet, 7% are using it to send email, and 2% to locate places using

GPS."[27] In Ofcom's May 2010 survey, Google Maps emerged as the seventh most popular site for the 25- to 34-, 35- to 49-, and 50- to 64-year-old demographic groups (whereas the top three sites for all age groups, in order, were Google, Google Search, and MSN/Windows Live/Bing).[28] As well as the use of GPS and Google Maps, location-aware and context-sensitive applications have now become popular in a number of countries. One country where these have been emerging has been the US, where Pew Internet research undertaken in late 2010 found that 4 percent of online adults used location-based services such as Foursquare or Gowalla, with 1 percent of Internet users using these on any given day.[29] A higher proportion of users who access the Internet via mobile (7 percent) or wireless technologies (5 percent) access "geosocial" or location-based services, and the same was true of online adults aged 18–29 years (8 percent).[30] In research on location-based service in the US, the UK, Germany, Canada, and Japan released by Microsoft in early 2011, the countries that featured as early adopters were Japan and the US. The top services used were overwhelmingly bound up with everyday place negotiation: GPS (70 percent), weather alerts (46 percent), traffic updates (38 percent), restaurant information and reviews (38 percent), and locating the nearest convenience stores (36 percent).[31] Social networking (18 percent), gaming (10 percent), geo-tagging photos (6 percent) lagged behind, alongside enhanced 911 (9 percent).

While such research is indicative at best, reporting on a fluid market, and emergent media practices, it depicts three important things about location-based services: (a) the new kinds of location-based services based on Internet use, especially on mobiles and wireless devices—such as social networking, gaming, and geo-tagging—are steadily gaining users, (b) these uses—at the time of writing at least—are nonetheless still only the pursuit of a small minority of all mobile and Internet users, and (c) the most common location-based services are still those to do with navigation (sat nav) and traffic, weather and news, and wayfinding and locating services.

The broader developments that make the present generation of the location-based services have to do with the affordance of smartphones, and how these new forms of mobile media are being used. Pew Internet research undertaken in 2011 found that 35 percent of American adults own a smartphone (while 83 percent of US adults have a cell phone of some sort).[32] The research found that "urban and suburban residents are roughly twice as likely to own a smartphone as those living in rural areas, and employment status is also strongly correlated with smartphone ownership."[33] In an Australian survey of 18- to 30-year-olds undertaken by myself and Kate Crawford, conducted at approximately the same time as the Pew study, we found that 72 percent owned a smartphone and 97 percent of them used apps.[34] In this study, we found the kinds of apps used most frequently by 56 percent of users were social networking software.

This quickening diffusion of smartphones—whether Apple's iPhone, the many devices using Google's Android operating system, Nokia and

Microsoft Windows Mobile 7, and many others—and the apps stores that have accompanied them (Apple Apps store, Android Market, Nokia Ovi, etc.) have provided a platform for software developers to offer apps.[35] Many of these apps avail themselves of the location information gathered by GPS, mobile cellular networks, status and location updates, tagging, and other data trails that are based on the capabilities of mobiles and the Internet. The underlying capabilities of smartphones particularly are key to this new emphasis on location.

How smartphones—indeed a range of mobile media devices can be utilized to gather data—was dramatized in April 2011 where researchers Alasdair Allan and Pete Warden attracted worldwide attention and concern arising from their paper presented to the *Where Is 2.0* conference at Santa Clara, California. Allan and Warden highlighted the amount of location data routinely collected and stored on a typical iPhone 4 or 3G iPad, occasioning widespread concern.[36] Allan and Warden revealed that:

> Ever since iOS 4 [what Apple promotes as the "the world's most advanced mobile operating system"] arrived, your device has been storing a long list of locations and time stamps. We're not sure why Apple is gathering this data, but it's clearly intentional, as the database is being restored across backups, and even device migrations . . . What makes this issue worse is that the file is unencrypted and unprotected, and it's on any machine you've synched with your iOS device. It can also be easily accessed on the device itself if it falls into the wrong hands. Anybody with access to this file knows where you've been over the last year, since iOS 4 was released.[37]

In response, Apple provided a clarification worth quoting at length. The corporation admitted to a bug in its software, but defended the need to gather data in this way:

> The iPhone is not logging your location. Rather, it's maintaining a database of WiFi hotspots and cell towers around your current location, some of which may be located more than one hundred miles away from your iPhone, to help your iPhone rapidly and accurately calculate its location when requested. Calculating a phone's location using just GPS satellite data can take up to several minutes. iPhone can reduce this time to just a few seconds by using Wi-Fi hotspot and cell tower data to quickly find GPS satellites, and even triangulate its location using just Wi-Fi hotspot and cell tower data when GPS is not available (such as indoors or in basements). These calculations are performed live on the iPhone using a crowd-sourced database of Wi-Fi hotspot and cell tower data that is generated by tens of millions of iPhones sending the geo-tagged locations of nearby Wi-Fi hotspots and cell towers in an anonymous and encrypted form to Apple.[38]

Apple's detailed response is fascinating for its description of how it per-ceives and constructs the location of the user, through a polling of the various technologies and networks that surround her. Through this series of "crowd-sourced approximations," Apple marshals a database construc-tion of location and place. It relies on the individual user, and indeed the mass of iPhone users collectively contributing their location data. This is something that Apple has designed itself as a way to encode place, relying on the co-creation of this by users. It is something that Apple believes users will avail themselves of, presumably to be able to quickly use the many capabilities of the phone, its operating system, and the apps that marshal these. This is a second, explicit sense in which users collaborate—wittingly, or otherwise—in the encoding of place that smartphones entail. From this telling episode in smartphones and place, let us now return to the larger scene—and ecologies—of mobile media and place.

SOCIAL RELATIONS OF MOBILE-INFLECTED PLACE

As I have briefly outlined, the construction of place in mobile technology has changed markedly in recent years when it comes to the use of location information—what I have referred to as "encoding place." In many ways, the fundamentals of mobile technology have remained substantially the same. The design of cellular mobile networks, and their need to locate users in order to transmit and receive the signals and digitally coded informa-tion that enables voice telephony and mobile data communications is rela-tively well established. The addition of GPS to handsets meant that another method of collecting and analyzing data to determine handset location was added to, and overlaid, that of cellular mobile networks. While Bluetooth has not been integrated into this system of location, it also has played an important role. The complementarity of WiFi and cellular mobile networks has been represented by dual-mode handsets that enable easy switching between the two. As is evident in the Apple iPhone data *cause célèbre*, the relative ubiquity of WiFi networks (and mobile technologies now symbiotic relationship with them) can be harvested to help determine location. These layers of mobile and wireless networks, with their concomitant devices and applications, combine to create mobile media.

Indeed, global mobile media are increasingly forged at the many inter-sections being created between traditional cellular mobile telecommunica-tions networks, and the varieties of Internet—what elsewhere I have termed "global Internets."[39] In addition, these new forms of mobile Internet are conjoined with other infrastructures, especially networks of location, mapping, and sensing networks,[40] but also digital broadcasting networks (e.g., mobile television)[41] and transportation networks (e.g., those used by trains or cars).[42] Each of these networks addresses the user in distinctly different ways; and users interact with, domesticate, resist, and shape these technological systems in culturally specific ways—evidence of glocalization in action. Thus, we have a complex set of mobile media systems, still unfolding around the world.[43] Unsurprisingly, users themselves are very much in the process of grasping the affordances they create, and the new encoding of place they enact.

For my purposes here, I wish to highlight the complexity of the use of location information in these new kinds of mobile media. The point has been well-made by Gordon and de Souza e Silva that the growing availability of location-based information is especially important to how mobile media are unfolding, and what users and publics make of these possibilities.[44] These location-based mobiles—and indeed geomedia generally—are premised, and fundamentally rely on, the dialectics of information sharing.

There is a governing assumption that users will mostly be happy to share their location information. The construction of choice regarding such increasingly prevalent information is a theme in everyday use of mobile social location technologies such as Foursquare, revolving around the interpretation and consequences that follow from users opting out of revealing their location, and the consideration of the reasons for such withholding of information, and in which situations it is likely to occur.[45] The sensitivities around control of personal information and privacy have encouraged—or, in the cases of various companies, most notably Facebook—compelled privacy settings of some sort to be integrated into the design of interfaces. In such cases, where users have some control over the release, circulation, and marshalling of their personal information, different relationships toward the encoding of place flow from this. In many situations, however, users at superficial and deep levels of the mobile media ecology have little freedom when it comes to use of personal information, including new location-based information. It is no longer simply adequate to approach such questions from the perspective of privacy, though this certainly is vitally important. The overarching question is far more profound and pressing—and goes directly to the politics of the encoding of place, and how users figure in this.

We are familiar with movements toward open acess to, architectures of, and use of, information, including location information. Users have demonstrated that they appreciate the ability to freely represent place, through annotation, tagging, mash-ups, and other ways to draw on, textualize, and share location information. There are now a range of initiatives, technologies, and practices that are often conjoined to offer users the ability to encode place in their own manner. These include the geographic mark-up language, KLM (http://www.opengeospatial.org/standards/kml/), online map-building applications, publicly available, "open" application programming interfaces (APIs), not to mention mobile camera and photography practices intertwined with Internet-based mapping.[46] Such practices of user-driven encoding of place—especially prevalent in participatory urbanism[47]—have been emphasized in discussions of the geospatial web,[48] but have not been so common when it comes to the "geomobile" web—let alone, the complex configurations of location-acquired information associated with today's complex mobile media.[49]

Yet, the consideration of participatory culture in and through location-based mobile media and information is raising profound questions: What role do users have in encoding place through location technologies? What accounts and architectures of place are suggested via particular location technologies? And as various ensembles of location technology emerge in mobile media, how open are these partly intended, partly accidental assemblages to a diverse range of actors with a range of commercial and non-commercial, private and public, aims and objects?

The difficulty in part lies with the expectations associated with, and ideas governing, mobile technology. Openness, structures, and code that support user participation are believed by many to be defining characteristics of the "public Internet." We actually only have a narrow understanding of what the terms of this openness are, and the forms openness and participation take across different cultural, linguistic, and social settings which shape the Internet in different places.[50] Nonetheless, the push for openness in the geoweb partakes of this vision of an open, public Internet—which is widely believed to be endangered in the current climate of a "walled garden" of apps and relatively closed social media such as Facebook.

Yet, there is little discussion of which I am aware about the social lives of information in the vast universe comprising GPS devices, not least the installed base of sat navs. Sat navs bring together great databases of information, in the form of maps and street directories, with location information available through GPS. Obviously, the compilers and publishers of street directories have been strong, commercial interests involved in partnerships in the sat nav industry—something critical to them because of the real potential of the new technologies to wipe the street directory off the map (so to speak). Yet, sat navs themselves have remained a relatively closed device, regarding which users have little say—nor are there interfaces that allow users to modify, reconfigure, or recombine sat navs. It may seem more than a little utopian, or fanciful, to muse upon the openness or otherwise of sat navs. Why it becomes a live issue is bound up with the ways in which various technologies of location are presently recombined in mobile media.

CONCLUSION

In the past, citizens and consumers may have had little stake in the oldest databases of all in telephony—telephone directories; other than whether or not they wished to have their phone number listed, to be available for other subscribers or interested parties. The link between the telephone number and the address created location information, that placed a subscriber. This link is potentially of rich significance, something always understood by historians[51] but which become evident in the rise of reverse phone directory services in the 1980s and 1990s.[52] In the present day, the amount and type of location information associated with mobile media has greatly proliferated, well beyond the trajectory envisaged from fixed telephony into mobile telecommunications.

In this chapter, I have traced the contours of location-information we now find in mobile media. Such information is the property of many distinct but now interlinked infrastructures, with applications that bridge these—something especially evident in the apps platforms of smartphones. I have argued that these dynamics are involved in a complex encoding of place—the social relations of which we still know little about. What is evident, however, is that there is a politics to this encoding, in which—like much of contemporary participatory digital culture—users are much invoked, and indeed relied on to contribute their data and digital traces for the vastly complicated, recursive, decentralized, and distributed systems of media to function.

Yet, the contribution of this compulsory "user-generated content" does not entitle particular users to many safeguards on, or share in the value of, the harvesting of this location-based mobile-created information; let alone a right to expect—or even imagine—mobile media infrastructures open for alternative practices of place-making with the new infinity of information. Of course, users are proceeding to make the place their own anyway, experimenting with unauthorized, unpredicted, and unprogrammed routines—outside the confines of the select options on offer (e.g., vying to be Mayor of Foursquare); and, in accordance with the ancient creativity and attachment to place and place-making, illuminated by place theory, that long precedes, and guides the domestication of, our epoch's mobile technologies.

NOTES

1. Lawrence Lessig, *Code and Other Laws of Cyberspace* (New York: Basic Books, 2000), and *Code: Version 2.0* (New York: Basic).
2. Alexander R. Galloway, *Protocol: How Control Exists after Decentralization* (Cambridge, MA: MIT Press, 2004); Alexander R. Galloway and Eugene Thacker, *The Exploit: A Theory of Networks* (Minneapolis, MN: University of Minnesota Press, 2007).
3. Mark B.N. Hansen, *Bodies in Code: Interfaces with Digital Media* (New York: Routledge, 2006).
4. Ravi Sundaram, *Pirate Modernity: Delhi's Media Urbanism* (London: Routledge, 2009).
5. Sarai, ed., *The Public Domain* (Delhi: Sarai and the New Media Initiative, CSDS; Amsterdam : Society for Old and New Media, 2001).
6. The new area of software studies brings a new focus to accounts of code: Matthew Fuller, ed., *Software Studies: A Lexicon* (Cambridge, MA: MIT Press, 2008). For broad stakes in debates over the values and politics of the

socio-technical, see Deborah G. Johnson and Jameson M. Wetmore, eds., *Technology and Society: Building our Sociotechnical Future* (Cambridge, MA: MIT, 2009).

7. Eric Gordon and Adriana de Silva e Souza, *Net Locality: Why Location Matters in a Networked World* (New York: John Wiley, 2011).

8. Adriana de Souza e Silva and Jordan Frith, *Mobile Interfaces in Public Spaces: Locational Privacy, Control, and Urban Sociability* (New York: Routledge, 2012).

9. Juha Korhonen, *Introduction to 3G Mobile Communications* (Boston: Artech, 2001), 355–365.

10. Federal Communications Commission, "Enhanced 9-1-1—Wireless Services," http://transition.fcc.gov/pshs/services/911-services/enhanced911/Welcome.html

11. Bardo Fraunholz, Chandana Unnithan, and Jürgen Jung, "Tracking and Tracing Applications of 3G for SMEs," in *Mobile and Wireless Systems Beyond 3G*, ed. Margherita Pagani (Hershey, PA: IRM Press), 131.

12. Fraunholz et al., "Tracking and Tracing," 131.

13. See: Jaap Haartsen, "Bluetooth: The Universal Radio Interface for Ad Hoc, Wireless Connectivity," *Ericsson Review* 75 (1998): 110–117; Henrik Arfwedson and Rob Sneddon, "Ericsson's Bluetooth Modules," *Ericsson Review* 76 (1999): 198–205.

14. S. Buttery and A. Sago, "Future Applications of Bluetooth," *BT Journal* 21 (2006): 48–55.

15. For instance, see Abdul Malik S. Al-Salman, "Broadcasting Commercial Advertising using Bluetooth Technology," *International Journal of Web Information Systems*, 2 (2006); 135–141.

16. See: Alice Crawford, "Taking Social Software to the Streets: Mobile Cocooning and the (An-)erotic City," *Journal of Urban Technology*, 15 (2008): 79–97; Lee Humphreys, "Mobile Devices and Social Networking," in *After the Mobile Phone? Social Changes and the Development of Mobile Communication*, ed. Maren Hartmann, Patrick, Rössler, and Joachim Höflich (Berlin: Frank & Timme), 115–130 and her "Mobile Social Networks and Social Practice: A Case Study of Dodgeball," *Journal of Computer-Mediated Communication*, 13.1, article 17, http://jcmc.indiana.edu/vol13/issue1/humphreys.html

17. *The Economist*, "The Connected Car," *The Economist*, June 6, 2009, EC.

18. TomTom, "MapQuest Partners with TomTom to Offer a Portable, Personal Navigation Device," Media release, September 27, 2005, http://corporate.tomtom.com/releasedetail.cfm?ReleaseID=319545.

19. *The Economist*, "The Connected Car."

20. Michael Rycroft, ed., *Satellite Navigation Systems: Policy, Commercial and Technical Interaction* (Dordrecht: Kluwer, 2003).

21. Lars Rasmussen, *From Australia to the World: The Story of Google Maps & Google Wave*, published public lecture (Sydney: Warren Centre for Advanced Engineering, University of Sydney, 2009).

22. Adam Pash, "Google Maps Multiple Destinations," *Lifehacker*, December 15, 2006, http://lifehacker.com/software/google-maps/google-maps-multiple-destinations-222183.php

23. See: Gretchen Wilkins, ed., *Distributed Urbanism: Cities after Google Earth* (London and New York: Routledge, 2010); and Vittoria Di Palma, "Zoom: Google Earth and Global Intimacy," in *Intimate Metropolis: Urban Subjects in the Modern City*, ed. Vittoria Di Palma, Diana Periton, and Marina Lathouri (New York: Routledge, 2009), 239–270

24. Wikipedia, "Google Maps," June 24, 2011, http://en.wikipedia.org/w/index.php?title=Google_Maps&oldid=435995442

25. Michael Arrington, "Google Redefines GPS Navigation Landscape: Google Maps Navigation For Android 2.0.," *TechCrunch*, October 28, 2009, http://techcrunch.com/2009/10/28/google-redefines-car-gps-navigation-google-maps-navigation-android/

26. Alice Crawford and Gerard Goggin, "Geomobile Web: Locative Technologies and Mobile Media," *Australian Journal of Communication* 36 (2009): 107.

27. Ofcom, *The Communications Market 2008*, August 14 (London: Ofcom, 2008; http://www.ofcom.org.uk/research/cm/cmr08/), 119.

28. Ofcom, *Communications Market Report 2010*, August 19 (London: Ofcom, 2010; http://stakeholders.ofcom.org.uk/market-data-research/market-data/communications-market-reports/cmr10/uk/), 267.

29. Pew Internet, *4% of Online Americans use Location-Based Services* (Washington, DC: Pew Internet, November 4, 2010), http://www.pewinternet.org/Reports/2010/Location-based-services.aspx

30. *Pew Internet, 4% of Online Americans use Location-Based Services*, 4.

31. Microsoft, *Location-Based Services are Poised for Growth*, January 2011, http://www.microsoft.com/download/en/details.aspx?id=3250.

32. Pew Internet, *Smartphone Adoption and Usage* (Washington, DC: Pew Internet, July 11, 2011).

33. Pew Internet, *Smartphone Adoption and Usage*, 3.

34. Survey conducted by Nielsen, in April–May 2011. Sample size was 1004 respondents. The research formed part of an Australian Research Council Discovery project *Young, Mobile, Connected*, discussed in Kate Crawford and Gerard Goggin, *The Politics of Mobile Social Media*, forthcoming.

35. Gerard Goggin, "Ubiquitous Apps: Politics of Openness in Global Mobile Cultures," *Digital Creativity* 22.3 (2011): 147–157.

36. Alasdair Allan and Pete Warden, "Got an iPhone or 3G iPad? Apple is recording your moves," *Radar*, 20 April 2011, http://radar.oreilly.com/2011/04/apple-location-tracking.html

37. Allan and Warden, "Got an iPhone or 3G iPad?"

38. Apple, "Apple Q & A on Location Data," http://www.apple.com/pr/library/2011/04/27Apple-Q-A-on-Location-Data.html

39. See: Gerard Goggin and Mark McLelland, ed., *Internationalizing Internet Studies: Beyond Anglophone Paradigms* (New York: Routledge, 2009); Gerard Goggin, "Global Internets: Media Research in the New World," in Handbook of Global Media Research, ed. Ingrid Volkmer (Malden, MA: Wiley Blackwell, 2011).

40. See: Dipankar Raychaudhuri and Mario Gerla, eds., *Emerging Wireless Technologies and the Future Mobile Internet* (New York: Cambridge University Press, 2011); and Christian van't Hof, Rinie van Est, and Floortje Daemen, *Check In / Check Out: The Public Space as an Internet of Things* (The Hague: Rathenau Institute; Rotterdam: NAi Publishers, 2011).

41. See: Rene de Renesse, *Mobile TV: Challenges and Opportunities Beyond 2011*, Open Society Foundation, June 2011, http://www.soros.org/initiatives/media/articles_publications/publications/mapping-digital-media-mobile-tv-20110627; and Gerard Goggin, "The Eccentric Career of Mobile Television," *International Journal of Digital Television* 2 (2011).

42. Gerard Goggin, "Driving the Internet: Mobile Internets, Cars, and the Social" *Future Internet* 4 (2012).

43. Gerard Goggin, *Global Mobile Media* (New York: Routledge, 2011).

44. Gordon and de Souza e Silva, *Net Location*.

45. de Souza e Silva and Jordan Frith, *Mobile Interfaces in Public Spaces*.

46. For instance, see Dong-Hoo Lee, "Re-imagining Urban Space: Mobility, Connectivity, and a Sense of Place," in *Mobile Technology: From Telecommunications to Media*, ed. Gerard Goggin and Larissa Hjorth (New York: Routledge, 2009), 233–249.

47. Malcolm McCullough, "On Urban Markup: Frames of Reference in Location Models for Participatory Urbanism," *Leonardo* 14, no. 3 (2006), http://leoalmanac.org/journal/vol_14/lea_v14_n03–04/mmccullough.html

48. Arno Scharl and Klaus Tochtermann, eds., *The Geospatial Web: How Geobrowsers, Social Software and the Web 2.0 are Shaping the Network Society* (London: Springer, 2007).

49. Note Teodor Milew's sophisticated, provocative critique of such "counter-cartographies," in which he suggests that:

> new media cartographies can only hope to unlock the enormous complexity of network topologies if they literally repopulate their maps with the rich performativities of subjects never detached from things. Counter-cartographies can be successful . . . only when they stop mistaking the effects for already-present contexts and instead concentrate on tracing how, through what transformations, detours, assemblies, and alliances are those effects produced. "Repopulating the Map: Why Subjects and Things are Never Alone," *Fibreculture Journal*, 13 [2008] http://journal.fibreculture.org/issue13/issue13_mitew_print.html

50. Goggin, *Global Mobile Media*.

51. Not least Claude S. Fischer in his *America Calling: A Social History of the Telephone to 1940* (Berkeley, CA: University of California Press, 1992).

52. Select Committee on Community Standards Relevant to the Supply of Services Utilising Electronic Technologies, *Report on 0055 Reverse Phone Directory Services* (Canberra: Parliament of Australia, 1993).

13 The Infosphere, the Geosphere, and the Mirror

The Geomedia-Based Normative Renegotiations of Body and Place

Francesco Lapenta

First there are the utopias. Utopias are sites with no real place. They are sites that have a general relation of direct or inverted analogy with the real space of Society. They present society itself in a perfected form, or else society turned upside down [. . .]. There are also, probably in every culture, in every civilization, real places—places that do exist and that are formed in the very founding of society which are something like counter-sites, a kind of effectively enacted utopia in which the real sites [. . .] are simultaneously represented, contested, and inverted. I believe that between utopias and these quite other sites, these heterotopias, there might be a sort of mixed, joint experience, which would be the mirror.[1]

INTRODUCTION: ABOUT TIME, SPACE, AND THE MEDIATED CONSTRUCTION OF PLACE

All life happens in space, and no action or idea can happen if not in place.[2] The history of place is the history of evolving moments in time in which society and individuals negotiate—by means of "cognitive,"[3] social,[4] and experiential interactions—the position of their bodies and the meanings of the objects that surround them.

Space becomes place when we cognitively question our own body, its limits and the realities that surround it, and also when we differentiate— "emotionally,"[5] physically, and experientially—between the objects space contains. Space becomes place when we transform or shape these objects according to our meanings, hence producing place, or when we socially negotiate these meanings and transform them into shared "values"[6] or social norms and conditions to interact among and with the objects place contains.

The history of place is the history of the efforts to communicate, share, or impose our own interpretations of place to others by means of negotiation, communication, and interaction or by means of coercion, imposition, or war. The history of place is also the history of continuous "movement" in place,[7] as we are always in the condition of moving (physically, interactively, or temporally) among ever-changing places. Hence, the history of place is the history of changing representations of place (cognitive, discursive, or mediated), as we maintain and communicate memories or meanings of past and present places to others or self.

The history of communication technologies is that of emancipation from space, and social renegotiation of place,[8] as technology is used to free ourselves from the limited space of existence, while maintaining ourselves in a technologically mediated projection (or representation) of our place of being. The history of mobile communication technologies is then just the latest evolution of these many histories of place, because it is just another history of cognitive, experiential, and social renegotiation or construction of place, rather than an impossible liberation or emancipation from place.

At the center of the latest evolution, pivot of social change, and cause of new forms of production and renegotiation of place, stands the mobile phone (or smartphone), or the new master of digital convergence that is the tablet. The introduction of the iPhone in 2007 and the iPad in 2010 constituted a cultural and technological shift that imprinted a decisive turn toward a novel interpretation of mobile communications and computing. Mobile phones have quickly moved from being simple, voice calling devices to being flexible, integrated, and personalized computational devices that offer scanning, multimedia, imaging, and other data-capture capabilities. The increasing personalization of mobile communication devices involved the transfer of established and highly personalized Internet communication tools and Web 2.0 applications—such as e-mail, Facebook, Twitter, YouTube, Skype, etc.—from the limited mobility of localized computing to the ubiquitous and continuous connectivity of mobile computing. It was the high mobility and continuous connectivity of mobile phones that fostered the latest evolution of mobile communication technologies that finally coupled available global positioning system (GPS) tools and new geographic information systems (GISs) to shape and support new location-based software and applications. Applications such as WikiMapia, OpenStreetMap, Google Earth or Google Latitude, Google Places, Foursquare, Gowalla, and Facebook Places, are but a few that use smartphones' newly acquired location-based interfaces and functionalities, and constitute a limited set of examples of existing and new Web 2.0 geomedia-based exchanges, communications, and interactions.

The evolution of mobile phones from simple communication devices into integrated and personalized multimedia centers of mediated communications has transformed the smartphone into the epicenter of contemporary mobile users' online existence—a personalized communication platform used to achieve "continuous connectivity"[9] with family, friends, and other social networks. The recent integration of new, location-based functionalities and applications has finally signaled the merging of what cannot be disconnected anymore: the online and offline identities and social worlds of contemporary media users.

It is my contention that these late technological shifts not only have implications for the ways we experience place and mediated social relations, but also for how mediated social relations can be theorized within a new *normative interpretation* of a media-based social reconstruction of place. Place is increasingly constructed, experienced, and consumed through new location-based communication technologies, or *geomedia*,[10] that challenge earlier evolutions of modern-day communication technologies. Analog communication technologies might have signaled the path for the emancipation of social interactions from the normative social dimensions of place and time. However, new location-based communications, I argue, are evidence of a new attempt to normatively, socially, economically, and cognitively redefine and reorganize place as a meaningful and socially regulated entity, quantity, and quality of mediated communications.

THE SPATIAL TURNS OF MEDIATED COMMUNICATIONS

Analog media worked as catalysts for an increasingly mediated and globalized society that challenged local normative definitions of place and interaction. Giddens, who was at the forefront of this interpretation, claimed that, because of the evolution of communication technologies, society evolved into a bipolar society characterized by two main modes of social interaction: one, *face-to-face*, remained "context bound" and dominated by "presence" and "localized" communications and exchanges and embedded in local definitions of place. The other, *mediated,* fostered by emerging communication technologies, was independent of place and local contexts and characterized by "relations between absent others."[11] This early stage, technologically induced, evolution of personal communications has fundamentally transformed modern social interactions, Giddens claims, creating the conditions for an increasingly globalized, mediated, and delocalized society. This concept was sustained and embraced by many media scholars who saw the Internet as further evolution of this placeless, mediated, and globalized society in which the Internet is interpreted as the "non-place," or virtual space, that conclusively "negates geometry"[12] and establishes the "death of distance"[13] in mediated social interactions.

These interpretations might initially have been justified by the historical trajectory and impact of analog media. Indeed, later the very unique nature of the Internet with its ubiquity, redefinition of space, and fragmentation of time and place, fundamentally questioned established definitions of social dimension such as "place" (hence the discussions of more abstract dimensions such as virtual space, virtual reality, virtual communities, and virtual identities) or "time." After all, most of what made the Internet unique was based on its ability to challenge modern definitions of time, space, and place, somewhat justifying the claims of "death of distance" and "negation of geometry." However, even in the excited times of the early digital era, there were calls for sustained attention to and evaluation of the contexts of what appeared context-less interactions. Goodchild reminded us that context and place are fundamental dimensions of social action.[14] Social action, no matter how small, abstract, or universal in its scope, always finds its roots in personal, social, and cultural definitions of time and place. A case in point is the very history of analog communications.

The progressive distanciation of communication technologies from a local, shared definition of place and time of social interactions started with the operational separation of time and space characteristic of modern communication technologies (e.g., radio, telegraph, and telephones). It increased with the development and wide-spread adoption of mobile communication technologies, and reached a new level of complexity with the evolution of new forms of mobile computing (e.g., e-mail, voice over Internet protocol [VOIP], short message service [SMS], multimedia messaging service [MMS], Tweets, etc.). This progressive evolution in communication technologies entailed a number of shifts in the definition and perception of the times and places of interpersonal communications, but also new and evolving negotiations of contexts, real or projected, socially shared or personally defined, and new social norms and practices that regulated them.

Mobile phones, for example, more than other communications technologies, brought about both a higher level of communicative flexibility (the ability to communicate in "any place," at "any time"), but also a higher level of socio-communicative unpredictability. The common shift from the "How are you?" of old to the more contemporary "Where are you?" of early mobile communications is an example of this separation of mediated communications from place, but also proof of the persistent social need to negotiate place as a defining social and communicative function in all mediated interactions. Meyrowitz made a timeless case for the *evolution* rather than *suspension* of established social categories such as time and place in all mediated communications, and argued that all new communication technologies should be interpreted as social sites of renegotiation of established social values and norms.[15] Hine, well aware of these preconditions, argued that even Internet-based communications and interactions root their meanings and contents in established contexts of production and experience, or patterns of cognitive interpretation that try to establish meanings and contents on assumed or projected definitions of place.[16] This quest for the definition of context has only been exacerbated by the recent evolutions of mobile communications and their increasing integration with these alternative forms of communications that have all but converged into this one mobile.

Despite these critical perspectives, however, the focus of digital media scholars and the public alike mostly remained focused on the evolution and social impact of the *placeless geometry of networks* that characterized the Internet. This was predictably and rightly so given the momentous evolutionary social function played by such placeless and often anonymous communications that remain revolutionary in nature, and evolving in scope. Yet, despite this mainstream attention to new theories of placeless networks or virtual communities and realities, a meaningful redefinition of the agenda of media scholars is occurring that has put renewed questions about space, time, and place at the forefront.

This chapter, informed by this emerging agenda, questions the validity of the old binary model initially described by Giddens, and, by building on Foucault's aging but never old allegory of the "mirror," argues that contemporary geomedia-based communications are better understood within a triadic model; a model that interprets contemporary smartphones and tablets and geomedia-based communications as a new territory, and new form of social interaction, neither place or virtual space, nor face-to-face or merely mediated communications, but rather a double sided "mirror" like relation, a third space of negotiation, in which the *infosphere* (the sphere of the sign)[17] and the *geosphere* (the sphere of the object and the body) reflect one another with different discursive and social effects on their utopian visualizations of place, and heterotopian actualization of place. The *Geosphere*: the sphere of the body and the object, the physical environments in which media users communicate and live. The *Infosphere:* the images, sounds, and texts, the bits of information, the iconic and symbolic representations of these physical environments that media users produce and share. Between them are the "mirrors" in the palms of our hands, the mobile phones and tablets with their graphic interfaces in which the geosphere and the infosphere digitally converge, at the same time "place" and "non-place," reflection of an heterotopian reality, or projection of utopian representations of self and place:

> The mirror is, after all, a utopia, since it is a placeless place. In the mirror, I see myself there where I am not, in an unreal, virtual space that opens up behind the surface [. . .]. But it is also a heterotopia in so far as the mirror does exist in reality, where it exerts a sort of counteraction on the position that I occupy [. . .] since I see myself over there. Starting from this gaze that is, as it were, directed toward me, from the ground of this virtual space that is on the other side of the glass, I come back toward myself; I begin again to direct my eyes toward myself and to reconstitute myself there where I am.[18]

The screens of our mobile communication devices, with the virtual maps and augmented reality interfaces they project, have come to embody a new social place, a third social dimension, a utopia, a "non place," as it exists only through the representations and visualization of other places. And an heterotopia, an actual feature of place that I can now touch and interact with, like with other objects that surround me; a new territory in which new cognitive, experiential, personal and social negotiation of place and value are simultaneously projected, represented, contested, and inverted moving back and forth from the infosphere to the geosphere each time reflected and transformed in the hybrid third space of the screen.

THE ALGORITHMIC TURN IN THE
REPRESENTATION OF BODY AND PLACE

My contention, based on interpretations developed in my previous work, that explores the social impact and evolution of these geomedia-based visual representations and interactions,[19] and Uricchio's definition of an "algorithmic turn"[20] in mediated representations, is that these technologies represent a significant shift from the socio-cultural trajectory of older communication technologies. In a world governed by a Durkheimian "socially shared" conception of time, the Universal Standard Time that cadenced our analog lives, mobile phones and early digital communication technologies represented an increasing emancipation toward a Kantian, "inner," interpretation of time; an evolution from the limits of "place," that meant the suspension of the social normativity of shared conceptions and interactions in place; an historical technological evolution that crystallized a polarization between the two main modes of social interaction (connoted by either the presence or the absence of the body and its contexts of existence). Mobile and early digital communication technologies have created a distinctive diversification between the normative systems that organize these two characteristic forms of social interactions. One, regulating face-to-face interactions and exchanges, is regulated by a socially elaborated and socially shared definition of time and place. And the other, mediated, characterized by personally elaborated, and interpersonally negotiated, definitions of time and immaterial place.

E-mail, text messaging, VOIP communications, and Web 2.0 platforms (Twitter, Skype, Google Talk, Microsoft Messenger, Facebook, MySpace, etc.), have all given origin to new communicative environments, all with their own medium/platform-specific redefinition of the times and spaces of interaction. These new communication technologies and platforms have, however, also created a set of independent norms and practices to regulate and organize their mediated social interactions. Each technology/platform encourages specific forms of exchange, but also devises internal symbolic/iconic/procedural codes and communicative norms that can be used to regulate communications and organize individual personal communicative environments (immaterial spaces) and networks. The different "personal status icons" in Skype, Messenger, or Google Talk, for example, replace the socially constrained and contextualized definitions of time and space, with a personal redefinition of personal time and personal (immaterial) space. Local time and individual physical social contexts are substituted by a simple iconic definition of "presence" or "absence" and various "degrees of availability" to communication. These systems replaced the set of expectations defined by (the Universal Standard) Time, and the social norms characteristic of the physical contexts of face-to-face communications (and early analog communications), with a set of idiosyncratic (platform specific) and personalized regulatory systems. The interactants use these symbolic/iconic/procedural codes to define their communicative status (personal time of being) and their symbolic "location" (social, contextual, discursive place of being) to regulate their interpersonal communications. I call the entirety of these symbolic systems, s*ocial navigation systems.*[21] These software applications, and many others, represent instances of what Uricchio calls the "algorithmic turn"[22] in photographic mapping, and what I define, paraphrasing Baudrillard, as the "fifth order of the simulacrum,"[23] an evolution of graphic forms and photographic representations that are transforming the once binary relation of the image with its objects of reference into a multitude of algorithmic reinterpretations each constituting "algorithmic interventions,"[24] or platform-specific renditions of time and space, all representing a "distanciation" from the social time and space of face-to-face interactions.

THE GEOMEDIA-BASED REACQUISITION
OF THE PLACE OF THE BODY

By bringing the body, place, and time (the three fundamental variables of surveillance) back together into mediated communications, new location-based communication technologies lead a counter-tendency in contemporary media evolution toward a new socio-normative relocalization of mediated interactions. This counter tendency represents an ontological and epistemological inversion in the separation, personalization, and interpersonal negotiation of the time and space of analog and early digital mediated communications. It exchanges algorithmic diversity (of time and space) with new "algorithmic regimes"[25] that use space as their common system of reference. This shift not only poses questions about the "purpose" of this inversion in technological evolution, but also poses questions about the "social function" of the normative qualities of this new geomedia-based reorganization of the once fragmented and personalized digital territory.

One interpretation could look at these new emerging ICTs, and the social dynamics they enable, as a form of empowerment of media users, or a natural evolution of Internet-based social practices and communications. It was after all only a question of time before the body would reacquire its centrality in the constant flow of information that propagates from and converges toward it. Media users constantly produce images, texts, and sounds that are exchanged, shared, and now increasingly tagged and geopositioned by new geomedia-based softwares and applications. The constant growth of these volunteered geographic information (VGI)[26] and geomedia-based systems and applications represents in many ways a simple evolution of

Web 1.0 and Web 2.0 media users' interest in utilizing the web to create, assemble, and share personal information about self with peers or the public. If we can argue that Web 1.0 can be interpreted as the initial move toward the simple transfer of content (image, text, sound) to a new medium and digital delivery system (the computer and the Internet), we can also argue that search engines like AltaVista, Yahoo!, and Google constituted the first organizational (and normative) tools that allowed us to navigate the immense archive of collected information and data (infosphere) available on the net. We can interpret Web 2.0 practices as an adaptation to the exponential growth of such data and an evolution that adopted richer and more personalized and socially relevant systems, such as personal social networks, to navigate, organize, and exchange information according to paradigms other than a text-based search string. In this new Web 2.0 ecosystem, social identity and social interactions are transformed into data,[27] and data become part and parcel of social identity construction,[28] "identity markers"[29] used as part of a process of identity elaboration, impression management, and self-presentation.[30] It is only a natural evolution of this trajectory to finally include the body, and its location, as another system of reference used to organize the available information in a fashion significant for the individual.

Geomedia and many software applications, such as Google Earth and Microsoft's Photosynth, seem to represent the response to this need for aggregation, redefinition, and evolution in the organizational criteria used to access data produced and collected by media users, who always exist in space and place, and whose bodies were disconnected from the flow of information they generated and consumed. In some respects, then, the evolution of geomedia systems (a synthesis of a technological definition of space, and graphic representation of place) seems evolutionary and perhaps represents a response to the need to technologically engage with a multisensorial[31] organization of the increasingly multimodal and mediated social interactions and data. In this perspective, the interpretation is simple; geomedia transform the geolocation of their users, their geosphere, into data, and connect these data to existing data, the sounds, texts, and images of the infosphere—like keywords used by search engines to organize and crawl the net in search of relevant information. The body's location and data pertaining to it (such as that provided by Web 2.0 applications) are coded and transformed into fine-tuned search strings for the selection and transfer of contextually relevant information (from and to the infosphere). In doing so, geomedia help users to navigate the infosphere in search of information relevant to their geosphere, and to systematically connect the information they produce, share, and exchange with other similarly contextualized world representations. Once a disconnected collection of images, data, and texts, the infosphere is now readily organized on a screen, synthesized and organized on a map, and technologically connected to a body and its coded social identity to facilitate the meaningful production and exchange of information with peers or the world.

LIFE ON A SCREEN: THE EVOLVING NORMATIVE RENEGOTIATIONS OF PLACE AND SELF ON THE VIRTUAL MAP

Another interpretation, however, could question what seems to go unseen in this interpretation, covered by these layers of mediated and localized representations and interactions, the "algorithms," the specifically designed graphic interfaces, the maps and projected layers, that at the same time support and transform the ways in which the user, the infosphere, and the geosphere reflect and interact with one another.

The contemporary evolution of location-based technologies, I have argued,[32] is characterized by two parallel evolutions. On the one hand, we have the increasing availability of advanced smartphones that integrate GPSs, augmented reality applications, scanning, multimedia, imaging, and other data-capture capabilities. This information and data constitute the building blocks of the new GISs that utilize "hardware, software, data, people, organizations, and institutional arrangements"[33] to collect, analyze, compile, visualize, and distribute data elaborated in the forms established by different geolocational platforms and applications, such as the virtual maps of Google Earth, the augmented realities of Layar, or the reconstituted images of Photosynth, which use space as their organizing principle. On the other hand, we have the contemporary growth and increasing availability of VGI, the texts, images, and sounds recorded and provided voluntarily by users of location-based Web 2.0 applications (WikiMapia, OpenStreetMap, Google Places and Google Latitude, Foursquare, Gowalla, MyMaps, etc.) that geo-tag and merge such data in live, virtual, and navigable maps or augmented reality interfaces.

These new virtual maps, I claim,[34] deserve in-depth scrutiny because of the complex social dynamics and developing social functions they have come to engender, and because of the significant technological and cultural evolution they represent both in terms of "representation" and "negotiation" of the time, space, and place of mediated interactions.

If the virtual map's origin and basic function lays in its capacity to facilitate and organize Geomedia users' collective production and exchange of images, sounds, and texts, it does so by visualizing information by means of elaborated algorithms that transform bits and pieces of information into a unified image—an augmented photographic map[35] enriched by layers of multimedia representations of the world. This photographic map, and its digital or "algorithmic"[36] transformation in the representation of the world, is leading a momentous technological, communicative, and cultural evolution. This is an evolution from the analog indexicality of the photograph of old, to the algorithmic "geolocality"[37] of the photosynthesis of the virtual map of today. This involves an evolution from a binary form of interaction (face-to-face, mediated) to a ternary (locative) alternative that combines elements of the two. It also entails a cultural evolution that is transforming the cognitive and experiential value of the map and the function that its virtual representations of space and place acquire in the re-elaboration and reorganization of the real spaces of existence, or real places of face-to-face interactions.

In the virtual map, the wealth of collectively produced information (i.e., VGI) is now readily synthesized and organized in a network of mediated social interactions and exchanges that are finally connected by geomedia technologies to what cannot be disconnected anymore, the individual's corporeal physical reality and the individual's coded digital identity. Because of its social, interactive, and localized nature, the virtual map may be conceived as a complex, evolving, interactive, and live visualization of the social identities, social relations, and social contexts of the users that contribute to its evolving representations. As a collective and social phenomenon, the geomedia-based virtual map has a transformative social significance: In 1967, Debord already described modernity as an immense accumulation of "spectacles," an immense collection of images of every aspect of life that fused in a common stream created "a pseudo-world apart."[38] He characterized the society of the "spectacle" not as a society simply "collecting images," but as a system of "social relations among people, mediated by images."[39] In 1991, Jameson described "cognitive mapping" as the coordination of existential data, the empirical position of the subject, with an abstract conception of an "unrepresentable" socio-geographic totality.[40] Cognitive mapping is a way of making sense of the world, of transforming space into place by cognitively, emotionally, physically, experientially, and socially questioning our own body, its limits, and the realities that surround it.

If framed within these theoretical perspectives, the virtual map can be interpreted as more than a mere collection of representations of the world; the virtual map can be interpreted as a new mediating social space—a projection of the individual and social mediations of the once disconnected "world of autonomous images" (Debord) and the "real" (Baudrillard),[41] "cognitive" and "social" worlds (Jameson) of their producers, a visible articulation of the individual and social inferences of the infosphere and the geosphere. The virtual maps and the augmented realities of our mobile devices can be interpreted as projective tools that transform our "cognitive mapping" of the world. They transform an abstract, personal, cognitive projection (our mental map of the world), into a collectively produced, and directly experienced and shared, virtual projection of the world. This can be conceptualized as a mirror, a reflection of a socio-geographic totality, which visualizes our individual and collective projections of the world, and our mediated conditions of existence. But it can also be conceptualized as a new social negotiation space in which personal representations, meanings, and values of place can be shared, agreed upon, or confronted, challenged, and inverted by other representations, meanings, and values.

The powerful projective, cognitive, and social potential of this new utopian and heterotopian space of interaction, poses questions about the "mimetic" nature of the algorithmic structures of the virtual map that *normatively* guide, and selectively represent, these exchanges and interactions on the map.

All the virtual maps, geomedia-based data aggregation tools, and augmented reality applications are, as all maps, "not neutral."[42] However, their bias is changing form and social function. Analog maps, by their own nature, are a "graphic reduction"[43] made evident by a graphically coded signifying system that always reveals its processes of reduction, and, most commonly, its tasks and specific communicative agendas (the coded graphics of subway maps being a very clear example). Virtual maps are a new generation of GISs that utilize "direct visualizations" or "visualizations without reduction."[44] These camouflage their reductions or "algorithmic interventions"[45] (as in the case of Google Maps's Photoshopping of military and sensitive sites) by relying on historically different cultural epistemologies and signification systems (such as the "photographic" that still embodies, although in a renegotiated form,[46] the "realist bias" of photography of old).

These algorithmic interventions can be interpreted as the idiosyncratic signatures of the different location-based applications. They can also, however, be interpreted as platform-specific "regulatory systems" that, by procedurally selecting among data and exchanges, not only superimpose organizational and institutional agendas that regulate and connote their signifiers, but, also, by selectivity transforming their virtual representations of place, transform and influence our "cognitive mapping" of place and space. The algorithmic structures of the virtual map rely on hardware, software, data, people, organizations, and institutions that embody different agendas in the designs and algorithms that collect, analyze, compile, synthesize, visualize, and distribute data elaborated for specifically designed tasks. If the specific agendas and tasks of each application and location-based software may not be generalized, we can, however, elaborate on what they all have in common and what their transformations underlay.

As these critical analyses have shown, the apparent promotion of flexibility and the autonomy enhancing qualities of new ICTs also come in constant tension with an opposite function that may allow us to interpret these communication technologies as new organizational and regulatory systems. In this perspective, geomedia not only can be interpreted as the evolution and response to the matured need for new organizational criteria to coordinate and link the flow of information in mediated interactions with the body. They can also be interpreted as the attempt of new actors, and the same economic and political forces that regulate physical time and place, to organize and regulate the global, and once placeless, flow of information into locally controlled, and physically contextualized geographic information systems. In these systems, not only are information linked back to their local referents (the physical place and the body of the user), but the user (and their surrounding place) is transformed into information, a commodified image, once again embedded in a controlled and socially and economically structured system regulated by new and old actors, and by new and old agendas. In this framework, geomedia and the virtual map can be conceived as the instruments enabling the capitalization, or control, of the new immaterial commodities and digital identities that dominate these new social environments (the utopian and heterotopian space in which these new forms of social interaction and exchange take place).

In my previous and forthcoming work,[47] I have focused on some economic dimensions of this transformation, and interpreted the geomedia-based social reorganization of information, place, and the body as the predictable expansion of the global market economies from the systematic organization and capitalization of time (sanctioned by the Universal Standard Time) to the systematic informatization and capitalization of individual's digital identities, immaterial places, and personal time. In that context, I claimed that location-based media or geomedia are to space what the watch is to time. Geomedia regulate social behavior, coordinate social interactions, and regulate interpersonal communications, while organizing the exchange of information, the founding immaterial commodities constitutive of these new immaterial places. I hereby conclude by suggesting that, as some considerations in this chapter might have shown, this is but one, if not the most evident, form of renegotiation that is affecting evolving normative renegotiations of place and self in these new geographically enhanced social media environments. The intrinsic control and algorithmic normativity of geomedia-based communications and exchanges pose questions, not only about surveillance and the normative power of its digital representations,[48] but also about its developing pervasiveness and social normativity.

Sanford Flaming's vision of a globally coordinated world time market, started with a system (Universal Standard Time), and the precise social scheduling of labor and capital; it eventually trickled down into the organization of the most mundane and personal events.[49] Geomedia are transforming the "virtual space" (the utopian space) of digital communications of old, in a mirror-like heterotopian "digital place," a projection that may come one day to directly influence and organize the real places in which we interact and live.

NOTES

1. Michel Foucault, "Of Other Spaces" (1967), in Nicholas Mirzoeff, ed., *The Visual Culture Reader* (London: Routledge, 2002), 231.
2. Thomas Hobbes, *Leviathan, or, The Matter, Forme, & Power of a Common-wealth Ecclesiasticall and Civill* (London: Andrew Crooke, 1651).

3. Fredric Jameson, *Postmodernism; or, The Cultural Logic of Late Capitalism* (London: Verso, 1991).
4. Michel Foucault, "Of Other Spaces," 229–237.
5. David M. Hummon, "Community Attachment: Local Sentiment and Sense of Place," in *Place Attachment*, ed. Irwin Altman and Setha M. Low (New York: Plenum Press, 1992), 262.
6. Yi-Fu Tuan, *Space and Place: The Perspectives of Experience* (St. Paul, MN: University of Minnesota Press, 1977), 6.
7. Nigel Thrift, "Movement-Space: The Changing Domain of Thinking Resulting from the Development of New Kinds of Spatial Awareness," *Economy and Society* 33 (2004): 582–604; Sarah Pink, "Sensory Digital Photography: Re-thinking 'Moving' and the Image," *Visual Studies* 26 (2011): 4–13.
8. Joshua Meyrowitz, *No Sense of Place: The Impact of Electronic Media on Social Behavior* (Oxford: Oxford University Press, 1985).
9. Jane Vincent, "Emotional Attachment," *Knowledge, Technology and Policy* 19, no. 1 (2006): 39.
10. See: Francesco Lapenta, ed., "Locative Media and the Digital Visualisation of Space, Place and Information," special Issue of *Visual Studies* 26, no. 1 (2011)—especially Lapenta, "Geomedia: On Location-Based Media, the Changing Status of Collective Image Production and the Emergence of Social Navigation Systems," *Visual Studies* 26, no. 1 (2011): 14–22; see also Francesco Lapenta, "GeoMedia: Looking Back at 2008 and the Early Developments of WEB 3.0," June 22, 2008, http://beingdigital.org/2008/06/22/geomedia-looking-back-at-2008-and-the-early-developements-of-web-30/.
11. Anthony Giddens, *The Consequences of Modernity* (Stanford, CA: Stanford University Press, 1990).
12. William J. Mitchell, *City of Bits: Space, Place, and the Infobahn* (Cambridge, MA: MIT Press), 8.
13. Frances Cairncross, *The Death of Distance: How the Communications Revolution Will Change Our Lives* (Boston: Harvard Business School Press, 1997).
14. See: Michael Goodchild, Luc Anselin, Richard P. Appelbaum, and Barbara Herr Harthorn, "Toward Spatially Integrated Social Science," *International Regional Science Review* 23 (2000): 139–159; Michael F. Goodchild and Donald G. Janelle, eds., *Spatially Integrated Social Science* (Oxford: Oxford University Press, 2004).
15. Meyrowitz, *No Sense of Place.*
16. Christine Hine, *Virtual Ethnography* (Thousand Oaks, CA: Sage, 2000).
17. See Alvin Toffler, *The Third Wave* (London: Bantam Books, 1980); Lapenta, "Geomedia."
18. Foucault, "Of Other Places," 232.
19. Lapenta, "Geomedia: On Location-Based Media," and "Geomedia: Looking Back at 2008."
20. William Uricchio, "The Algorithmic Turn: Photosynth, Augmented Reality and the Changing Implications of the Image," *Visual Studies* 26, no. 1 (2011): 25–35.
21. Lapenta, "Geomedia: On Location-Based Media."
22. Uricchio, "The Algorithmic Turn," 25.
23. Lapenta, "Geomedia: On Location-Based Media."
24. Uricchio, "The Algorithmic Turn."
25. Uricchio, "The Algorithmic Turn."
26. Michael F. Goodchild, "Citizens as Sensors: The World of Volunteered Geography," *GeoJournal* 69 (2007): 211–221.
27. Jenny Sundén, *Material Virtualities: Approaching Online Textual Embodiment* (New York: Peter Lang, 2003).

28. danah boyd, "The Significance of Social Software," in *BlogTalks Reloaded: Social Software Research & Cases*, ed. Thomas N. Burg and Jan Schmid (Norderstedt: Books on Demand), 15–30.

29. Richard West and Lynn H. Turner, *Understanding Interpersonal Communication: Making Choices in Changing Times* (Boston, MA: Cengage Learning), 389.

30. Judith Donath and danah boyd, "Public Displays of Connection," *BT Technology Journal* 22 (2004): 71–82.

31. Sarah Pink, "Sensory Digital Photography."

32. Francesco Lapenta, "Geomedia-Based Methods and Visual Research: Exploring the Theoretical Tenets of the Localization and Visualization of Mediated Social Relations with Direct Visualization Techniques," in *Advances in Visual Methodology*, ed. Sarah Pink (Thousand Oaks, CA: Sage, 2011).

33. Kenneth Dueker and Daniel Kjerne, *Multipurpose Cadastre: Terms and Definitions* (Falls Church, VA: American Society for Photogrammetry and Remote Sensing and American Congress on Surveying and Mapping, 1989).

34. Francesco Lapenta, "Geomedia: On Location-Based Media."

35. Francesco Lapenta, "Mapping the World. A Brief Essay on the Changing Status of Digital Photography," 2008, http://www.scribd.com/doc/9896688/Mapping-the-World-A-Brief-Essay-on-the-Changing-Status-of-Collective-Image-Production-F-Lapenta.

36. Uricchio, "The Algorithmic Turn."

37. Lapenta, "Geomedia: On Location-Based Media."

38. Guy Debord, *Society of the Spectacle* (London: Rebel Press, 1983 (1967)).

39. Debord, *Society of the Spectacle*.

40. Jameson, *Postmodernism*, 52.

41. Jean Baudrillard, *Selected Writings*, ed. Mark Poster (Stanford, CA: Stanford University Press, 1988).

42. Rob Walker, "The Lie of the Land: Mark Monmonier on Maps, Technology and Social Change," *Visual Studies* 26, no. 1 (2011): 62–70.

43. Lev Manovich, "What is Visualisation?," *Poetess Archive Journal*, http://paj.muohio.edu/paj/index.php/paj/article/view/19/58.

44. Lev Manovich, "What is Visualisation?"

45. Uricchio, "The Algorithmic Turn."

46. Lapenta, "Geomedia: On Locative Media," 17.

47. Francesco Lapenta and Jakob Arnoldi, "Geomedia and the Capitalization of Virtual Space," forthcoming.

48. Media scholars would call that just "business as usual."

49. Theodor W. Adorno, *The Culture Industry: Selected Essays on Mass Culture*, ed. J. M. Bernstein (London: Routledge).

Contributors

Caroline Bassett is Reader in Digital Media and Director of the Centre for Material Digital Culture at the University of Sussex. Her work explores digital technology and cultural forms and practices. She is currently working on noise and anti-noise, digital humanities as critical theory, and is completing a monograph on anti-computing.

Chris Brennan-Horley holds a PhD in Human Geography from the University of Wollongong, Australia. His research interests cover the intersection between Geographic Information Systems (GIS) and cultural research. His dissertation forged new understandings of the creative city by combining mental maps with GIS methods. He has published articles on creative work, festivals, and spatial aspects of the creative city.

Edward S. Casey is Distinguished Professor at SUNY, Stony Brook, where he is an active member of the PhD program, widely known for its focus on continental philosophy. He is also co-director of a new master's program in Manhattan that is entitled Philosophy and the Arts. He has studied and taught in France and Germany, where his basic training was in phenomenology. His books include *Imagining, Remembering, Spirit and Soul: Essays in Philosophical Psychology, Getting Back into Place, The Fate of Place, Representing Place in Landscape Painting and Maps,* and *Earth-Mapping.* More recently, he has concentrated on the importance of the glance in human visual experience in a new book *The World at a Glance.* In a volume to be entitled *The World on Edge,* he is now engaged in a study of edges as they figure into geographical, linguistic, political, and aesthetic contexts.

Richard Ek has a PhD in human geography and is Associate Professor at the Department of Service Management, Lund University. His research interests include areas such as the history and philosophy of spatial thought, critical geopolitics, political theology, strategic communication, and place branding.

Yoriko Inada is a researcher at the Department of Social Sciences at the Télécom Paristech (Ecole Nationale Supérieure des Télécommunications, France). She has a PhD in ethnology from the University of Paris. Her recent studies focus on the hybridization of digital and physical environments.

Christian Licoppe is Professor of Sociology and head of the social science department at Telecom Paristech. Trained in sociology of science and technology for several years he has been studying the organization of interactions and activities mediated by information and communication technologies. In the particular field of mobile studies he has introduced the notion of "connected presence" and conducted (with Yoriko Inada) one of the first extensive studies of a location aware community of mobile players. He is currently interested in the co-shaping of locative media and mobile user behavior.

Jeff Malpas is Professor of Philosophy and Australian Research Council Professorial Fellow at the University of Tasmania and Distinguished Professor of Philosophy at La Trobe University. He is the author of *Heidegger and the Place of Thinking* (MIT Press, 2011), *Heidegger's Topology* (MIT Press, 2006), *Place and Experience* (Cambridge University Press, 1999), and of numerous articles in journals and collections. He is best known for his work on the philosophy of place and space, hermeneutics and philosophy of language, ethics, transcendental philosophy, and twentieth-century German and American thought. He is regularly engaged in interdisciplinary work across a range of disciplines including art, architecture, geography and social theory, and also engages with philosophy in the public sphere commenting on ethical and political issues in a range of media. He is a former Humboldt Fellow and serves on the editorial boards of a number of journals.

Chris Gibson is Professorial Fellow, Australian Research Council Future Fellow and Deputy Director at the Australian Centre for Cultural Environmental Research, University of Wollongong, Australia. His research interests span the fields of human geography, cultural planning, and creative arts. He is the author of *Sound Tracks: Popular Music, Identity and Place* (Routledge, 2003) and *Music Festivals and Regional Development* (Ashgate, 2010). His current collaborative research projects are on festivals, cultural asset mapping, and the cultural economy of household sustainability.

Eric Gordon is an Associate Professor in the Department of Visual and Media Arts at Emerson College. His research focuses on location-based media, mediated urbanism, and games for civic learning. He is the director of the Engagement Game Lab at Emerson College where he designs and studies games and social media to enhance urban civic engagement. His game Participatory Chinatown was named "Best Direct Action Game" by Games for Change in 2011. He is the co-author, with Adriana de Souza e Silva, of *Net Locality: Why Location Matters in a Networked World* (Blackwell, 2011), and the author of *The Urban Spectator: American Concept-cities from Kodak to Google* (Dartmouth College Press, 2010). His recent work has been published in journals such as *New Media and Society, Information, Communication and Society, Environment and Planning B, Space and Culture*, and others.

Gerard Goggin is Professor of Media and Communications, Department of Media and Communications, the University of Sydney, Australia. His books include *New Technologies and the Media* (2012), *Global Mobile Media* (2011), *Internationalizing Internet Studies* (with Mark McLelland), *Mobile Phone Cultures* (2008), *Cell Phone Culture* (2006), *Virtual Nation: The Internet in Australia* (2004), *Digital Disability* (with Christopher Newell), as well as the *Routledge Companion to Mobile Media* (2013) and *Mobile Technologies: From Telecommunications to Media* (2009), both with Larissa Hjorth. Current research includes a number of Australian Research Council-funded projects on: youth and mobile media; Internet history in Asia-Pacific; distribution of audio-visual material across traditional, mobile, and online media.

Larissa Hjorth is artist, digital ethnographer and Senior Lecturer in the Games and Digital Art Programs at RMIT University. Since 2000, Hjorth has been researching on gendered customizing of mobile communication, new media literacy, gaming and virtual communities in the Asia–Pacific—these studies are outlined in her book, *Mobile Media in the Asia-Pacific* (London, Routledge, 2009). Hjorth has published widely on the topic in national and International journals in journals such as *Games and Culture, Convergence, Journal of Intercultural Studies, Continuum, ACCESS, Fibreculture* and *Southern Review,* and recently co-edited two Routledge anthologies, *Games of Locality: Gaming Cultures in the Asia-Pacific* (with Dean Chan) and *Mobile Technologies: From Telecommunication to Media* (with Gerard Goggin). In 2007, Hjorth co-convened the International *Mobile Media* conference with Gerard Goggin (www.mobilemedia2007.net) and the *Interactive Entertainment* (IE) conference with Esther Milne (www.ie.rmit.edu.au). In 2009 she began her Australian Research Council discovery fellowship (with Michael Arnold) exploring the role of the local and online within communities in the Asia-Pacific region. This three year cross-cultural case study will focus on six locations—Tokyo, Seoul, Shanghai, Singapore, Manila, and Melbourne.

Francesco Lapenta is Associate Professor in Visual Culture and New Media at the Department of Communication, Business and Information

Technologies, at the University of Roskilde. He is a member of the editorial board of the journal *Visual Studies*, a member of the executive board of the Intentional Visual Sociology Association and currently a Visiting Professor at the Sociology Department of New York University. Lapenta's most recent work includes the special issue "Autonomy and Creative Labour" of the *Journal for Cultural Research* (July 2010, and the article "Geomedia: on Location-Based Media, the Changing Status of Collective Image Production and the Emergence of Social Navigation Systems" (2011), and the special issue of *Visual Studies,* "Locative Media and the Digital Visualisation of Space, Place and Information" (March 2011).

Susan Luckman is Portfolio Leader: Research and a Senior Lecturer in the School of Communication, International Studies and Languages at the University of South Australia who teaches and researches in the fields of media and cultural studies. She is a foundation member of the ARC Cultural Research Network; author of the forthcoming monograph *Place, Affect and Cultural Industries: The Politics and Poetics of Creative Work* (Palgrave Macmillan); co-edited the anthology on creative music cultures and the global economy (*Sonic Synergies*, Ashgate, 2008); and is the author of numerous book chapters, peer-reviewed journal articles and government reports on creative cultures and industries. Susan is an interdisciplinary cultural studies scholar whose work is concerned with how place is an agent in cultural practice. Using mostly interview and ethnographic methodologies, her research has explored the intersections of place and culture in relation to dance music cultures, new media use, creative and cultural industries, the socio-political effects of media representation, and attitudes to higher education participation.

Ingrid Richardson is a Senior Lecturer in the Faculty of Creative Technologies and Media at Murdoch University, Western Australia. Her broader research interests include philosophy of science and technology, new and interactive media theory, phenomenology, visual ethnography, and embodied interaction. She has published journal articles and book chapters on the cultural and corporeal effects of mobile media, virtual reality, biomedical imaging, technologies for sustainability, TV, and urban screens.

Adriana de Souza e Silva is an Associate Professor at the Department of Communication at North Carolina State University (NCSU), affiliated faculty at the Digital Games Research Center, and a faculty member of the Communication, Rhetoric, and Digital Media (CRDM) program at NCSU. Dr de Souza e Silva's research focuses on how mobile and locative interfaces shape people's interactions with public spaces and create new forms of sociability. She teaches classes on mobile technologies, location-based games and Internet studies. Dr de Souza e Silva is the co-editor (with Daniel M. Sutko) of the book *Digital Cityscapes—Merging Digital and Urban Playspaces* (Peter Lang, 2009), and the co-author (with Eric Gordon) of the forthcoming book *Net-Locality: Why Location Matters in a Networked World* (Blackwell, 2011). She holds a PhD in Communication and Culture from the Federal University of Rio de Janeiro, Brazil.

Iain Sutherland is a developer, writer, and lecturer in new media technologies. He completed a PhD at the University of Melbourne in 2010 and has published in a range of journals on social, phenomenological, and technological consequences of ubiquitous mobile communications. Iain currently lectures at the University of Melbourne and develops applications which leverage networked digital technologies in local environments.

Rowan Wilken is a lecturer in Media and Communications, Swinburne University of Technology, Melbourne, Australia. He is the author of *Teletechnologies, Place, and Community* (Routledge, 2011), as well as numerous articles that examine the relationship between place and media. His present research interests include digital technologies and culture, domestic technology consumption, mobile and locative media, old and new media, and theories and practices of everyday life.

Index

A

action at a distance 179
Actor-Network Theory (ANT) 39,
    43–9, 160; ANT 2.0 44; ANT
    and after 44; citizen-mobile 44;
    citizen-weapon 44
algorithmic turn 218–19, 221–4
Allan, Alasdair, 205
AltaVista 220
always-on 115
Amin, Ash 42
analog media 215
Apple 27; retention of iPhone location
    records 199–200, 205–6
apps 4
at-homeness 159, 170
attention: mobiles and 14; *see also*
    everyday distraction, presence
auditory space, *see* soundscape
Augé, Marc 4, 110; non-places 10–12,
    110
autonomy 180

B

ba ling hou generation 140–53; impor-
    tance of mobile media to 152–3;
    narrative of place 147
Bachelard, Gaston 90
Backhaus, Gary 193
Barcelona 159, 162–70
Bassett, Caroline 11, 14
Baudrillard, Jean 222
Beck, Ulrich 105
being in place 159, 170
Berland, Jody 5
Blanchot, Maurice 6, 112, 113
blogs 35
Bluetooth 64, 68, 198, 200–1
bodily knowledge 164

body: as material-semiotic assemblage
    41; routine 15; subject 158, 169
body–metaphors 181
body–screen relation 181
body–screen–place relation 188, 194
body–technology relation 15, 41, 180,
    183, 189, 194: phenomenologi-
    cal accounts 13–15
body–tool relation 41, 180, 182–3
Bourdieu, Pierre 112–13, 183; *see also*
    habitus
Bowles, Kate 131–2
brain-training games 104–16; as
    self-renovation projects 107,
    112–14
Brightkite 199
Brown, Steve 44, 45
Bull, Michael 114, 192

C

Cartesian: space, 43, 44–5, 50; map
    126
Casey, Edward S. 4, 175–80, 181,
    183–4, 185, 186, 189
Castells, Manuel 4, 8, 40, 90
cellular mobile networks 205
chat rooms 98
Chesher, Chris 45–6
China Education and Research Net-
    work (CERNET) (China) 143
China: cross-generational media
    literacy 143; generational differ-
    ences in media use 152; impact
    of one-child policy 147, 148;
    internet usage rates 144; mobile
    media in 140–53; new media
    policy in, 148; role of social
    media 144–7; rural internet
    usage 144; *see also ba ling hou*

generation, Electronic Information Service System (EISS) (China) project
citizen journalism 4
citizen-mobile 44
citizen-weapon 44
cognitive mapping 222
co-location work 78
communication: computer-mediated 16; as information transmission 179–80; location of 16
computer-supported collaborative work (CSCW) 57, 58, 66
connection in-place 106
connectivity, 33; enhanced 34–5; individual 26–7, 33–4
Connor, Steven 193
conviviality 114–15
Cooley, Heidi Rae 14–15, 189; screenic seeing 15
co-presence 16, 41, 62–4, 73, 89, 140, 141, 142, 151, 185, 193
corporeal schemata 183
Coyne, Richard 13
Crawford, Alice 203
Crawford, Kate 184
Creative Nation 135
Cresswell, Tim 4, 5–6, 181, 191
Cronon, Bill 124–5
cultural geography 9
cultural globalization, 7, 8, 9

D

Darwin (Australia): broadband speed 129; *Committee for Darwin Report* 135; links to remote communities 128–9; mobile phone reception 129; technological mediation of place in 125, 126–36
de Certeau, Michel 4; conception of space and place 10–11
de Souza e Silva, Adriana 17, 90, 184, 189–191, 199
Debord, Guy 222
Deleuze, Gilles 45
Denigan, Michael 131
Derrida, Jacques 188–9
diaspora: role of mobile technology 141
digital: adaptation 131–5; broadcasting 205–6; chatter 35; cityscapes 58, *see also* public space; divide 135; self 91

disembedding technologies 158, 160, 166–7
Dodgeball application (US) 201
Doel, Marcus 42
Dourish, Paul 17, 90
Dragon Quest 9 57–86; Akihabara 59, 67, 70–3; in France 79–83; in Japan 60–2, 70–3; Odaiba Fuji TV 59; paired players 69; as proximity game 60–2; Shizuoka Station 59, 74–7; social media and 66–7; *see also* proximity games
Dunn, Kevin 123
Durkheim, Emile 218
DVD players, portable 27
dwelling 40

E

Electronic Information Service System (EISS) (China) project 143, 148, 152
email 35, 152, 203
embodied existence: impact of mobile technologies 175–80
embodied location 192–3, 195
embodiment relation 183, 191
e-readers 4
Euclidian space 47
everyday: banality 6; distraction 14, 32; movement 163
evidential: boundaries 66, 84; insideness 159, 170

F

Facebook 79, 208, 214, 218; ban in China 144; privacy settings 207
Facebook Places 214
Farman, Jason 191–2
Fetion 144
field composition 191
Flaming, Sanford 224
flows 8, 40
Foucault, Michel 105
Foursquare 95, 99, 199, 204, 214, 221
frame 181; as lived logic 186; as organ of perception 186; as window on the world 186
freezing of being 179
Friedberg, Anne 12, 17, 181, 186–7
Friedman, Susan 9
Frith, Jordan 189, 199
front and back stages 161
Fudan University 143, 148

G

Galloway, Alexander 199
gaming: adventure 107; aesthetic
    qualities 106; location-based
    191; puzzle 107–8, *see also*
    brain-training games; structure
    of game 112–13; *see also* micro-
    gaming, *Dragon Quest* 9
Garmin 201–2
Geocaching 191–2
geographical: end of geography thesis
    126, 134; impact on technology
    126–36; physical qualities 125;
    space 123, 125; theories of place
    123–36
geography 9; *see also* cultural geogra-
    phy, humanist geography
geolocality 221–2
geomedia 215; possibilities for cre-
    ation 219; regulation of social
    behavior 224; transformation of
    virtual space 224
geomobile web 203, 208; openness in
    208
geospatial web applications 4, 184,
    198, 199, 203, 208, 219; *see
    also* Google Maps
geosphere 217
geo-tagging 204, 221
Gergen, Kenneth 16, 89; absent pres-
    ence 16, 89
Giddens, Anthony 105, 215
Gilroy, Paul 114–15
GIS systems 130
GISs 214, 221
globalization 7, 27, 33–4, 35; *see also*
    cultural globalization
glocalization 8–9, 185, 207
Goffman, Erving 63, 66, 73, 91, 94–5,
    97, 99, 108–9, 114, 161; front
    and back stages 151; going away
    96–7, 100; production of the
    self 108; typologies 68, 76
Goggin, Gerard 45, 123, 184
going away 96–7, 100
Goodchild, Michael 216
Google 220
Google Android 204–5
Google Earth 198, 199, 214, 220, 221
Google Latitude 214, 221
Google Maps 4, 126, 198, 199, 202,
    204, 223; navigation app
    202; *see also* geospatial web
    applications

Google Places 214, 221
Google Talk 218–19
Gordon, Eric 17, 92, 199
Gowalla 204, 214, 221
GPRS 144
GPS navigation 125, 126, 142, 176–7,
    199, 201–2, 214, 221; illegal
    aliens and 177; in mobile phones
    201
Green, Nicola 114

H

habitudinal edge phenomena 186
habitus 113, 183–4; renovation of 114
Hansen, Mark 199
Harrison, Steve 90
Harvey, David 90, 115–16; telepathic
    city 115–16
Heidegger, Martin 40–1, 183, 193
Helyer, Nigel 193
here–there relation 181
Hjorth, Larissa 9, 16, 125
Höflich, Joachim 14
home: relocation of domestic ties 141;
    mobile technology and 147–8;
    sense of 140–1
homeomorphism 46
Hong Kong: Filipino workers in 141
horizontality 33
humanist geography 4, 157
human–mobile assemblage topologies
    39–51; *see also* social topologies
hybrid cultural ecology: of internet
    café 57, 58; of public spaces
    57–86
hybrid engagements with place 17,
    79–83
hypothetical imperative 178

I

iBranding 27
identity performance 108
Ihde, Don 182–3; embodiment relation
    183, 191
Ilharco, Fernando 183, 187, 188, 189
individual: as primary unit of connec-
    tivity 33
individualization of place 26–36
information as place 90
information management 178
information-sharing, dialectics of 207
infosphere 217; embodiment 220–1; on
    map 220; on screen 220;
interconnection 27

International Telecommunications Union 3
Internet: access in Australia 125–7; café 57; mobile 4; as non-place 215; as placeless geometry of networks 216; public 208; social practices 219; varieties of 205–6
interstitial zones, 11, 72, 215; *see also* non-places
Introna, Lucas 183, 187–189
IP address 68
iPad 3, 27, 190, 205, 214
iPhone 3, 27, 45, 190, 204, 205, 214; location data stored on 205, 221; *see also* smartphones
iPod 94, 192, 194; creation of sonorous envelope 192
Ito, Mizuko 10, 141

J
Jameson, Fredric 222
Jobs, Steve 190
just-in-time 115
Juul, Jesper 107

K
Kaixin 144; demographics of use 146
Kant, Immanuel 178, 188, 218
Katz, James 15; choreography of mobile communication 169
Klein, Norman 89
KLM mark-up language 207
Kopytoff, Igor 142
Koryakukan forum 67

L
lag 113
LamdaMOO 98
Lapenta, Francesco 17
Latitude 199
Latour, Bruno 43
Laurier, Eric 16, 165
Law, John 39, 44–8
Layar 221
Lazarus principle 110
Lee, Nick 44, 45
Lefebvre, Henri 90, 108, 110–11; concept of space 90
Lessig, Lawrence 198–9
Ling, Rich 97, 161
local space 14
localization, of human life 27–8

location: impact on social media use 126–36, 149, 151; pervasiveness in mobile technology 126, 136
locational turn 198
location-aware: mobile media 184, 189–92; technologies 91, 100
location-based: gaming 191; social networks 58; *see also* locative media, geo-tagging
locative media 124, 198, 199, 204, 214; history 200–3
Loopt 93, 95, 199
LoveGety device (Japan) 201
Lukermann, Fred 5
Luminosity.com 104, 110, 111

M
Mace, Michael 190
Malpas, Jeff 4
Manovich, Lev 187, 188
Massey, Doreen 4, 9, 70, 140, 141–2
material objects: biographies of 142
Matsuda, Misa 141
McKay, Deirdre 141
McLuhan, Marshall 40
mediated encounters 60–1, 62–4; behaviour patterns 63; brave engagements 75–7; hybrid encounters 64; mediated apparitions 64; minimal hospitality 63; negative face 63; personal territories 63; random encounters 64–5; right to tranquility 63; timid encounters 75–9, 84; *see also surechigai tsushin*
membership categorization analysis 69
memory and place 110–12
Merleau-Ponty, Maurice 166–7, 182; freezing of being 179
metropolis 93
Meyrowitz, Joshua 4, 8, 13, 16–17, 27–8, 32, 36, 90, 216
micro-gaming 105–7; triviality of 106–7
Microsoft Messenger 218–19
Microsoft Photosynth 220, 221
migrant workers: location awareness, sociality and 57–8; technology and 141, 145;
Miller, Vince 105–6
Misa, Thomas 4
Mixi forum 67
MMS 216

mobile communications: choreography of 15, 169; colonization of urban space 194–5; third nature of 194

mobile phone: 3G networks 3; 4G networks 3; communicative flexibility of 216; as containers of information 185–6, 188; coverage in rural areas 3; distraction and 14, 32; evolution of 45, 214; as extension of body 30, 186; as external repository of memory 31; form 45; function 45; impact on place 4–5; impact on soundscape 192–4; international coverage 3; location and 165; location-based functionality 214; materiality of 31; net-local 189, 190–1; place in relation to humans 30; self-projection 34; social disruption and 89; social etiquette 162; as status symbol 31; subscriber numbers 3, 160; in Tanzania 48–9; ubiquity of 3; usage time 35; use in public 14, 32, 161; users' bodily gestures 161;

mobile privatization 140, 161; geography and 136; impact on embodied existence 175–80; as interface to public space 199; ordering of world by 35–6; politics of 198–209; as projection of world 36; significance of place and 140; *see also* technology, mobile phone

mobile technology: history of 214;

mobility: impact on place 7, 28; notions of 140

modified we-relationship 193

Mol, Annemarie 39, 44–8, 50

Moores, Shaun 90

MOOs 98

Morley, David 141

Morse, Margaret 14

movement: between places 26; in place 26

MP3 players 27, 114; creation of sonorous envelope 192

MSN in China 144

MUDS 98

MyMaps 221

MySpace 218; in China 144

net locality 89–100, 189, 190–1, 199; new contexts for interaction 90–1

N

netbooks 27

network coverage 125

networked: connectivity 89; individualism 92–3; locality, *see* net locality; society 40

networks 205–6

news, mobile 4

non-places 10–12, 215; internet as 215; screens as 217–18

O

Ofcom 203

ontology: relational 41; renegotiation 40; shift 42; spatial 49–50

open access to information 207

open APIs 207

OpenStreetMap 214, 221

P

parents in the pocket 142

parergon 181, 188–9

Paris 59; Jardin des Plantes 59; Parc de Bercy 59

participatory urbanism 207–8

PDAs 27, 190, 194

Perec, Georges 106–7, 108, 111–12

performance: in public space 98–100

personal privacy 207

Pfaff, Julia 48, 49

phenomenological geography 157, 159, 166

phenomenology 4; post-phenomenology 182, 184, 194; *see also* phenomenological geography

phone space 14, 161

place: changing experience of 28; concept of 4, 5–7; definitions of 90; degeneration of 157; discontinuities of 31; as dynamic psycho-social construct 90, 123; embodied engagement with 12; encoding of 17, 198; as engines of capitalism 124; erasure of 7; as event 42; history of 213–14; human lives in 29–30; impact of mobilities 7; impact of technology on 158; as information 90; liberation from 28, 33, 35, 36;

materiality 5; meaning 5; meaningful 90; mediated construction of 213; mobile-inflected 206–8; mundane practice 6; networked 90; phenomenological construction of 157–70; physical qualities 125; and placelessness 166; practice 5–6; as 'proper, stable and distinct location' 124–5; recombinations of 12–13; relational 9–10, 41, 42–3; screen as 184; sensory engagement with 12; social 90; soundscape of 192–4; textures of 13–17; as time-space compression 124; topology of 43–9; transformation in to position 35; uniqueness of 157; *see also* individualization of place, movement, sense of place

place ballet 169, 191
Poster, Mark 12
presence 15–17, 193; absent 16, 218–19; degrees of availability 218–19; distant 193; located 185; mediated 176; in place 177; in Web 2.0 applications 218–19; *see also* co-presence, tele-presence
production of the self 108–9
proximity games 59, 60–2, 69, 84; behaviour patterns 63; culture and sociality 83–6; in France 79–83; in Japan, 60–2, 70–3; spatial distribution of play 62; *see also* mediated encounters, *surechigai tsushin*
proximity: desire for 192; role in shaping experience of place 125;
public space: colonization by mobile technology 194; *Dragon Quest 9* in 62–7, 68–9; etiquette 95–6; gaming in 62–7, 68–9, 114–15; good 98; as hybrid ecologies 58, 190; mediation by technology 91–3; mobile phone use in 14, 32, 161, 194–5; performance in 98–100; playful 84; public–private space division 108–9, 114; renegotiation of being in 190–1; shifting behavioural norms 95–6; sociality in 68–9; *see also* proximity games

Q
QQ 142, 143, 144, 145–7, 149, 151; chat function 151; co-presence and 151; demographics of use 145–6; factors impacting success 149–50; impact of familiarity 150–1; video service 148

R
regulation of information 223–4
relationality 27, 33–4; networks of relations 33; of place 34; quantitative 34; spatialized 34
Relph, Edward 4, 5, 7, 157; place and placelessness 166
remoteness: technology and 126–36
Renren 142, 144; demographics of use 146
Rheingold, Howard 89
Richardson, Ingrid 14, 39, 41, 43, 48
Robertson, Roland 8–9
Rosenberger, Robert 191

S
Sanders, Rickie 126
Sarai 199
satellite navigation (sat nav) technology 198, 201–2, 204, 208
screen: boundary work of 186; as frame 186–7; as non-place 217–18; as place holder 186, 194; as window 186–7
screenic seeing 15
screen-ness 183, 187–8
screen–window relation 181–2
screen–window–frame relation 182
seamfulness 66, 68, 76, 84
seamlessness 66, 68
Seamon, David 4, 7–8, 13–14, 15, 17, 157–8, 163, 166–7; bodily knowledge 164; place ballet 169, 191; pre-conscious body subject 169
search engines 220
selective sociality 93
self: connection with place 29; externalized conception of 30; mobile phone as extension of 30, 31–2; sense of 29
Selzer, Mark 109
sense of place 5, 14, 15, 35, 124, 127, 140, 141
Serres, André 45